THE ENVIRONMENTAL IMPACT OF CITIES

The Environmental Impact of Cities assesses the environmental impact that comes from cities and their inhabitants, demonstrating that our current political and economic systems are not environmentally sustainable because they are designed for endless growth in a system which is finite.

It is already well documented that political, economic and social forces are capable of shaping cities and their expansion, retraction, gentrification, re-population, industrialisation or de-industrialisation. However, the links between these political and economic forces and the environmental impact they have on urban areas have yet to be numerically presented. As a result, it is not clear how our cities are affecting the environment, meaning it is currently impossible to relate their economic, political and social systems to their environmental performance. This book examines a broad selection of cities covering a wide range of political systems, geography, cultural backgrounds and population size. The environmental impact of the selected cities is calculated using both ecological footprint and carbon emissions, two of the most extensively available indices for measuring environmental impact. The results are then considered in terms of political, economic and social factors to ascertain the degree to which these factors are helping or hindering the reduction of the environmental impact of humans.

This book will be of great interest to students and scholars of sustainability, urban planning, urban design, environmental sciences, geography and sociology.

Fabricio Chicca is a senior lecturer in the School of Architecture, Victoria University of Wellington, New Zealand.

Brenda Vale is a professorial research fellow in the School of Architecture, Victoria University of Wellington, New Zealand.

Robert Vale is a retired professorial research fellow in the School of Architecture, Victoria University of Wellington, New Zealand.

ROUTLEDGE STUDIES IN SUSTAINABILITY

Circular Cities
A Revolution in Urban Sustainability
Jo Williams

Stories of Change and Sustainability in the Arctic Regions
The Interdependence of Local and Global
Edited by Rita Sørly, Tony Ghaye and Bård Kårtveit

Sustainable Living at the Centre for Alternative Technology
Radical Ideas and Practical Solutions
Stephen Jacobs

Rural Governance in the UK
Towards a Sustainable and Equitable Society
Edited by Adrienne Attorp, Sean Heron and Ruth McAreavey

The Environmental Impact of Cities
Death by Democracy and Capitalism
Fabricio Chicca, Brenda Vale and Robert Vale

For more information on this series, please visit: www.routledge.com/Routledge-Studies-in-Sustainability/book-series/RSSTY

THE ENVIRONMENTAL IMPACT OF CITIES

Death by Democracy and Capitalism

Fabricio Chicca, Brenda Vale and Robert Vale

Routledge
Taylor & Francis Group
LONDON AND NEW YORK

earthscan
from Routledge

Designed cover image: Fabricio Chicca

First published 2023
by Routledge
4 Park Square, Milton Park, Abingdon, Oxon OX14 4RN

and by Routledge
605 Third Avenue, New York, NY 10158

Routledge is an imprint of the Taylor & Francis Group, an informa business

British Library Cataloguing-in-Publication Data
A catalogue record for this book is available from the British Library

Library of Congress Cataloging-in-Publication Data
Names: Chicca, Fabricio, author. | Vale, Brenda, author. |
Vale, Robert (Robert James Dennis) author.
Title: The environmental impact of cities : death by democracy and capitalism /
Fabricio Chicca, Brenda Vale and Robert Vale.
Description: 1 Edition. | New York, NY : Routledge, 2023. |
Series: Routledge studies in sustainability |
Includes bibliographical references and index.
Identifiers: LCCN 2022033147 (print) | LCCN 2022033148 (ebook) |
ISBN 9780367493431 (hardback) | ISBN 9780367493424 (paperback) |
ISBN 9781003045847 (ebook)
Subjects: LCSH: City planning–Environmental aspects. |
Cities and towns–Growth–Environmental aspects.
Classification: LCC HT166 .C4425 2023 (print) |
LCC HT166 (ebook) | DDC 307.1/216–dc23/eng/20221017
LC record available at https://lccn.loc.gov/2022033147
LC ebook record available at https://lccn.loc.gov/2022033148

ISBN: 978-0-367-49343-1 (hbk)
ISBN: 978-0-367-49342-4 (pbk)
ISBN: 978-1-003-04584-7 (ebk)

DOI: 10.4324/9781003045847

Typeset in Bembo
by Newgen Publishing UK

CONTENTS

FIGURES

TABLES

ACKNOWLEDGEMENTS

The authors would like to thank following interns who tirelessly looked for data for this project, with special thanks to Lea Hungrichhaußen (Germany) for her two internships and assistance with Chapter 8.

Fabiola Wimmer (Germany)
Lea Medina (the United States)
Hannah Bendull (the United States)
Tin Sai Lun (China)
Karina Iurkova (Russia)
Mingjie Qiu (China)

Thanks also go to the following who also searched for data.

Lucas Felix Castro (Brazil)
Miguel Arroyo (Argentina)
George Lominnack (the United States)
Karen Green (Australia)
Michelle Lennox (UK)

PREFACE

This book was written to investigate the extent to which the rising global urbanisation would affect "earth-overshoot", the fact that humanity is consuming the earth's resources faster than they are being replenished.

Living in cities, especially in high-density environments, has been claimed by some people as being more sustainable, because less land is occupied by each person, people live in smaller dwellings, and being much closer together they also use much less fossil-fuelled transport (Breheny, 1992; Jenks et al., 1996). David Owen (2004), a writer for *The New Yorker*, made a claim for the sustainable credentials of living in cities when he described his family's move from an apartment in Manhattan to the suburbs as "an ecological catastrophe", going on to note the family's use of electricity "went from roughly four thousand kilowatt-hours a year, toward the end of our time in New York, to almost thirty thousand kilowatt-hours in 2003—and our house doesn't even have central air-conditioning." At a personal level, living in the city centre, being able to walk to work and to leisure activities, and losing heat mostly to neighbouring apartments, seems as if it would save energy.

However, the personal consumption of resources is not the only mechanism at work within cities, as this book will show. Where studies have been carried out, the ecological footprint (EF) of a city inhabitant is higher than the footprint for a regional or national average citizen (Chicca et al, 2018:3), suggesting those living in cities have a greater impact on the planet than non-urban dwellers. The idea for the research behind this book was to look at a range of different cities in differing climates and with differing types of governance to see if some ways of urban living are more impactful than others.

That was the aim, but it soon became very clear that what we wanted to do was almost impossible because of the lack of data. This is problematic for those espousing a move to sustainability—how can we see a path to where we want to be if we do not know where we are?

The first part of this book, therefore, deals with the struggle to find data in order to make some type of comparison of the ecological impacts of our selected cities. It also includes an investigation of the sustainability policies of the cities and how these relate to their ecological footprints (EFs). How we have done the calculations is also explained. The second part looks at attributes of cities, such as cultural values and tourism, population growth, capitalism and democracy, and how these might affect the impact of city dwellers. The book concludes with some sobering thoughts on what a low-impact city might be like.

Although this was not the book we set out to write, it has shown how very, very much some human ways of living have to change if humanity is ever to live within the resources of the planet.

References

Breheny, M.J. (1992). *Sustainable Development and Urban Form*. London: Pion.

Chicca, F., Vale, B. and Vale, R. (2018). *Everyday Lifestyles and Sustainability*. Abingdon: Routledge.

Jenks, M., Burton, E. and Williams, K. (1996). *The Compact City. A Sustainable Urban Form*. London: Routledge.

Owen, D. (2004). Green Manhattan. *The New Yorker*. Retrieved from www.newyorker.com/magazine/2004/10/18/green-manhattan, 1 April 2022.

PART 1

The impact of cities

1

THE CITIES

1.1 Introduction

City is a word that conveys a shared meaning but that is also hard to define clearly. We all think we know what a city is, but as Rozenfeld et al. (2011) noted, "A key difficulty in studying cities is finding a practical way to define them". Cities come in many shapes and sizes and with different forms of administration. The common characteristic of a city is a place where many people choose, or are forced by circumstances, to live close together, but this is not the only thing that makes a city; otherwise any refugee camp would be recognised as a city, indeed some are "... virtual cities in view of their population and demographic density" (Montclos and Kagwanja, 2000).

The United Nations Refugee Agency (UNHCR, n.d.) perhaps gives a clear definition of a city by noting "Most of the world's refugee camps were designed as temporary facilities. However, many have grown and developed into fully fledged cities, replete with vibrant economies, systems of governance, and civic institutions." Thus the city is the place where people come together to create things that can be traded, building an economy while putting in place methods to control this activity and to organise matters so that life in such close proximity is bearable.

From this viewpoint, a city would seem to be a good thing, since it allows people a means of living independent of the land on which it is based, while a rural society is bound to the land. Herein also lies the issue with cities, their separation from the natural world as suggested by E.M. Forster in *Howards End*, where he describes expanding London as nature withdrawing (Forster, 1910. (1941):102).

Of course, nature is still evident in cities, if only in the seasons, the pigeons, the rats and the cockroaches, but the nature of all cities is that they rely on land from elsewhere to supply the needs of their citizens, whether for food, clothes or building materials. This is clear in looking at the environmental impact, as measured

DOI: 10.4324/9781003045847-2

by the ecological footprint (EF) of those living in cities, since the average EF of city dwellers is higher than the national average where the city is located. For a full discussion of EF and how it is calculated, see Chapter 2. Hubacek et al. (2009) compared the EF of China to that of Beijing. They calculated the EF of China to be 1.78 global hectares for each Chinese citizen (gha/person) while that of a citizen of Beijing was over 80% greater at 4.99gha/person. In simple terms, a global hectare is a hectare of land of global average productivity to make comparisons between places possible. (A fuller description of how a global hectare is calculated can be found elsewhere (Chicca et al., 2018:126–128).) Obviously, a city has to expend energy on bringing in resources to support the lifestyle of its citizens and to dispose of their wastes, so its EF would tend to be higher than the EFs of those living closer to the land.

The difference between urban and rural footprints suggests other forces are at work, which raise the average environmental impact of each citizen. A second study that used a different calculation method calculated the EF of Beijing would double between 1996 and 2020, while over the same time the GDP would be 19 times greater (Lui and Lei, 2018), suggesting that the economy of a city and the consequent generation of wealth has a great effect on its environmental impact.

The problem is that more people are predicted to live in cities, with their consequent higher EFs, at a time when humanity is already using up the resources of the planet faster than they are being replenished (GFN, 2021a). Globally in 2018, 55% of people were urban dwellers, a figure predicted to rise to 68% in 2025 (UN, 2018). The global rural population dropped from just over 66% of total in 1960 to just under 45% in 2018 (The World Bank Group, 2019).

At the same time where and how people live within the same city can differ considerably. "Nearly all of the world's large urban areas have extensive suburbs of much lower density outside the historic cores that are characterized by higher densities" (Demographia, 2020:9). In terms of environmental impact is it better to live at high densities or lower densities within a city, or is it just city living that is the environmental issue? Whether the continued concentration of people into cities will be a problem for the future of the planet is the focus of this book, initially by examination of the current environmental impact of 19 cities in different continents and climates.

1.2 The cities and their history

How to introduce and group the 19 cities selected for study in this book is a problem. Our seemingly rather random selection was initially influenced by the apparent availability of data, something which has not been as apparent as first hoped. The aim was also to have a spread of cities with different characteristics, such as size, location, climate, as well as indicators like gross domestic product (GDP) and human development index (HDI). These characteristics are briefly covered in this first chapter, beginning with the history of the cities and where they are located.

Cities, as suggested above, occur because of trade, which is why many cities grew up by the sea or around rivers, though the geography of the river could dictate how the city grew. As an example, the city of Cairo is 1,000 years old and was originally situated to the east of the River Nile at the point where the river valley emerges from being constrained by the desert escarpments and before it begins to expand into the Nile delta. The oldest parts were to the east, away from the river's edge and its flood plain, but since then the city has grown westward as well as north and south, with further expansion on the western banks of the Nile (AlSayyad, 2021). Table 1.1 sets out some of the history of the chosen cities, from oldest to youngest, including where and when they grew up and, where this information is known, the main goods that made the city famous. The table is ordered by the date of each city's establishment. This table was compiled from the entries on cities in Encylopaedia Britannica (2021) unless otherwise stated. including where and when they grew up and, where this information is known, the main goods that made the city famous. The table is ordered by the date of each city's establishment.

TABLE 1.1 City origins

City	*Date of establishment*	*Geography*	*Traded goods*
Beijing	1045 BC[1]	On a plain between Yongding and Chaobai rivers	Established as the imperial capital so a centre of consumption rather than production
Pyongyang	108 BCE	On the Taedong River 30 miles inland	Established as trading point with China
Cairo	969 AD	On east bank of Nile where it widens to delta	Became the capital
Copenhagen[2]	1043 (first written record)	On islands at southern end of Øresund	Herring fishery trade
Moscow	1147	On Moskva River	Became capital Grand Duchy Moscow 13th century
Hamburg[3]	1189 (Free Imperial City)	By lower reaches River Elbe	Trading power
Amsterdam[4]	c1200	Around dam on Amstel River	Fishing village
Singapore[5]	14th century (modern Singapore 19th century)	Singapore Island	Trading post
Ahmedabad	1411	Along the eastern side Sabarmati River	Cotton (19th century onwards)
Havana	1519	Natural deep-water port	Colonial city

(continued)

TABLE 1.1 Cont.

City	*Date of establishment*	*Geography*	*Traded goods*
Buenos Aires	1536	Where Paraná River delta widens becoming the Río de la Plata estuary	Port for the export of agricultural products
São Paolo	1554	On a plateau 70k from Atlantic Ocean	Developed in 17th century when gold was discovered, followed by sugar cane growing
Jakarta[6]	1619	At mouth of Ciliwung River	Dutch colonial capital city engaged in spice trade
New York[7]	1624	On Hudson River	Colonial trading post
Sydney	1788	Around Port Jackson harbour	Colonial city
Dubai	1799	On Dubai Creek on southern shore Persian Gulf	Became capital in 1971 when UAE formed - previously centre of fishing, pearl fishing, and port
Auckland[8]	1841	Isthmus between Manukau and Waitemata harbours	Colonial capital until 1856
Kuala Lumpur[9]	1857	Where Gombak and Klang Rivers join	Tin mining
Nairobi	1899	On railway line from Mombasa to Uganda	Colonial city

1 TravelChinaGuide (2021).
2 Copenhagen.com (2020).
3 Hamburg.com (n.d.).
4 Amsterdam.info (2021).
5 VisitSingapore.com (n.d.).
6 UNESCO (2021).
7 Ferguson (2021).
8 Wilson (2016).
9 Gillick (2018).

Table 1.1 shows some groupings. Apart from the three very old cities of Beijing, Pyongyang and Cairo, the European cities were all established in the early Medieval period and, apart from Copenhagen, all were founded on rivers for ease of moving goods. Another group have similar dates of establishment in the 16th century linked to colonisation of South America and the Caribbean. Later colonisation led to the establishment of Sydney and Auckland in the Antipodes and Nairobi in Africa. Only two of the 19 cities were not linked to river or seaports, these being São Paulo, which was established by Jesuits on an inland plateau and Nairobi, the youngest city of the 19, which was established on the railway. The importance of trade in the history of cities is very evident.

1.3 Population

Another characteristic of cities is their population. This would seem to be a simple matter of establishing the boundary of the city and counting up all the people living within that boundary but it is not that easy. Table 1.2 groups the cities by continent with the 2021 population according to one source (World Population Review, 2021) and gives a ranking by population from 1 (the greatest population) to 19.

Exactly where the boundary of each city is drawn turns out to be highly problematic. For example, Hugo et al. (2003) show that the criteria for city boundaries in the United States are revised periodically. The concept of the "Metropolitan Statistical Area" was developed for the 1980 US census, and comprises "...a county or counties that contained either a city of at least 50,000 people (using legal boundaries to define the city) or an urbanised area of 50,000 or more population, with the total metropolitan population exceeding 100,000." This rather confusing definition was an attempt to recognise that the life of those living near the city in the adjacent counties is bound up with that of the "real" city dwellers in both social and economic terms. Inevitably under this definition, the area defined will change over time as populations change.

TABLE 1.2 The 19 cities and their populations as defined by World Population Review (2021)

City	*Continent*	*Population 2021*	*Population ranking*
Cairo	Africa	21,322,750	2
Nairobi	Africa	4,922,192	12
Ahmedabad	Asia	8,253,226	7
Beijing	Asia	20,896,820	3
Dubai	Asia	2,921,376	14
Jakarta	Asia	10,915,364	6
Kuala Lumpur	Asia	8,210,745	9
Pyongyang	Asia	3,108,108	13
Singapore	Asia	5,991,801	10
Amsterdam	Europe	1,157,519	19
Copenhagen	Europe	1,358,608	18
Hamburg	Europe	1,788,995	16
Moscow	Europe	12,593,252	5
Havana	N. America	2,142,866	15
New York	N. America	8,230,290	8
Auckland	Oceania	1,630,092	17
Sydney	Oceania	4,991,654	11
Buenos Aires	S. America	15,257,673	4
São Paulo	S. America	22,237,472	1

Cities, their areas and their populations are defined differently by different sources. Table 1.3 sets out population, area and population density of built-up urban areas as defined by Demographia (2020:2). In their terms, "An urban area is best thought of as the 'urban footprint'— the lighted area ('city lights') that can be observed from an airplane (or satellite) on a clear night." The 19 cities are also ranked by population density.

TABLE 1.3 The 19 cities, population data and ranking by population density from Demographia (2020:22–92)

City	*Continent*	*Population 2021*	*Area covered 2020 (square kilometres)*	*Population density*	*Population density ranking*
Cairo	Africa	19,372,000	2,010	9,639	5
Nairobi	Africa	6,020,000	851	7,065	7
Ahmedabad	Asia	7,410,000	359	20,583	1
Beijing	Asia	19,372,000	4,172	4,658	10
Dubai	Asia	6,036,000	1,508	4,004	11
Jakarta	Asia	34,540,000	3,540	9,756	4
Kuala Lumpur	Asia	8,285,000	2,162	3,831	12
Pyongyang	Asia	2,228,000	176	12,651	2
Singapore	Asia	5,745,000	523	10,981	3
Amsterdam	Europe	1,140,000	316	3,608	13
Copenhagen	Europe	1,321,000	570	2,187	17
Hamburg	Europe	1,976,000	794	2,485	16
Moscow	Europe	17,125,000	5,891	2,908	14
Havana	N. America	1,610,000	319	5,054	8
New York	N. America	20,870,000*	12,093	1,700	19
Auckland	Oceania	1,333,000	524	2,548	15
Sydney	Oceania	4,578,000	2,036	2,123	18
Buenos Aires	S. America	16,157,000	3,221	5,015	9
São Paulo	S. America	22,046,000	3,166	7,076	6

* New York-New Jersey-Connecticut.

These two tables demonstrate the problem of comparing statistics from different sources. There are big differences in the populations of Jakarta (11 million in Table 1.2 and 35 million in Table 1.3) and New York (8 million in Table 1.2 and 21 million in Table 1.3). What has been measured in Table 1.3 is the population of Greater Jakarta, which is made up of Jakarta and the surrounding areas of Bogor, Depok, Tangerang, Bakasi, Puncak and Cianjur (Winarso, 2011:163–164). The population density of just Jakarta, sometimes known as Jakarta's Special Capital Region, is somewhat higher than that of Greater Jakarta at 16,262 persons per square kilometre (population approximately 10.8 million and area 662.33 square kilometres) (World's Capital Cities, 2021). This population density would rank Jakarta second in Table 1.3 after Ahmadabad.

A similar situation accounts for the difference for New York. The United States Census Bureau (n.d.) gives the July 2019 population figures as 8,336,817 for New York city, New York and 19,453,561 for New York. The New York metropolitan area consists of New York city and the surrounding counties. The July 2015 population density of New York city (City of New York, 2021) is given as 27,000 persons/square mile (10,385/sq.km), which would make it fifth in the rankings in Table 1.3.

The discrepancies for Moscow are also due to the differences between the 2020 population of Moscow 12.5 million and Greater Moscow 20 million (World Population, 2020). The same source gives the area of Moscow as 2,511 sq.km and that of Greater Moscow as 26,000 sq.km. This makes the population density of Moscow 4,978 persons/sq.km, which would lift its ranking to 11th in Table 1.3, with a similar population density to Havana and Buenos Aires. Table 1.4 sets out the revised population densities from highest to lowest using just the central city values for Jakarta, New York and Moscow (italicised in Table 1.4).

TABLE 1.4 Population density ranking using central city values

City	*Persons / sq.km*	*Ranking*
Ahmedabad	20,583	1 (unchanged)
Jakarta	16,262	2
Pyongyang	12,651	3
Singapore	10,981	4
New York	10,385	5
Cairo	9,639	6
São Paulo	7,076	7
Nairobi	7,065	8
Havana	5,054	9
Buenos Aires	5,015	10
Moscow	4,978	11
Beijing	4,658	12
Dubai	4,004	13
Kuala Lumpur	3,831	14
Amsterdam	3,608	15
Auckland	2,548	16
Hamburg	2,485	17
Copenhagen	2,187	18
Sydney	2,123	19

Table 1.4 shows some clusters, such as the very similar space per person in São Paulo and Nairobi (around 7,000 people per km^2) and Havana, Buenos Aires and Moscow (around 5,000), with the four cities with the most space per person forming two groups of Auckland and Hamburg (2,500) and Copenhagen and Sydney (2,100). Obviously, these figures hide very different distributions of space for many citizens of each city. For example, in Nairobi 32% of the population were

estimated to live in slums, amounting to approximately 1 million citizens occupying 6% of the residential land in the city (Goudet et al., 2017). This figure may well be larger now, being based on a 2008–09 health survey and the population is now greater. Slums are also a characteristic of São Paulo, but here slums do not necessarily mean the same thing as in Nairobi.

> … they represent a big problem for the municipality; nevertheless, it is useful to stress that their qualitative state cannot be compared to slums in African or south Asian cities. São Paulo's favelas and cortiços have basic facilities that are quite unthinkable in the shanty towns of Lagos or Mumbai; 98.4 per cent of the buildings in the metropolitan area of São Paulo are connected to water, 70 per cent of the favelados and the inhabitants of cortiços are legally connected to the electricity distribution network, gas is used by everyone as fuel for cooking, and the collection of waste is guaranteed all over the city.
>
> *Chiodelli, n.d.*

These issues are explored further in Chapter 7. What this very short explanation of the population of the 19 cities has shown is the difficulty of finding data, in this case the problem of knowing where the city boundary is drawn, and hence how many people live inside it.

1.4 Climate

In a city, climate affects not just daily life but also how much energy the citizens might use and the ease with which green spaces can be maintained within the urban area. For example, trees can provide useful shade in hot climates, but if that climate is dry as well as hot, it may be difficult to find the water necessary to maintain the trees. In colder climates buildings that are contiguous can help to reduce the need for heating.

Various systems for classifying climate have been suggested, among which the Köppen-Geiger system is widely used. Stemming from its introduction in the late 19th century by the German botanist Wladimir Köppen, the system comprises five main climate zones, which can then be sub-divided, all based on temperature and what can grow (National Geographic, 2021). The five zones are A (tropical), B (dry), C (temperate), D (continental) and E (polar). All 19 cities fall into the A–D zones. To fit in zone A, the minimum temperature must be greater than or equal to 18°C. To fit into zone B, the annual precipitation must be less than 10 times the precipitation threshold or dryness index (this is the balance between precipitation and evaporation and will vary according to when the rain falls). To be in zone C, the minimum temperature must be greater than –3°C and less than 18°C. To fit into zone D, the minimum temperature must be less than or equal to –3°C (Kottek et al., 2006). Table 1.5 gives the climate zones of the 19 cities from Climate-Data.org (n.d.). The fourth and fifth columns give the annual average temperature and rainfall from the same source.

TABLE 1.5 The cities and climate

City	*Zone*	*Description*	*Average temperature (°C)*	*Annual rainfall (mm)*
Havana	Aw	Tropical savanna	25.7	821
Kuala Lumpur	Af	Tropical rainforest	25.8	2981
Jakarta	Am	Tropical monsoon	26.4	2097
Singapore	Af	Tropical rainforest	26.7	2366
Ahmedabad	BSh	Hot semi-arid	27.1	757
Cairo	BWh	Hot desert	22.1	18
Dubai	BWh	Hot desert	28.2	68
Buenos Aires	Cfa	Humid subtropical	17.3	1112
Amsterdam	Cfb	Oceanic	10.7	844
Auckland	Cfb	Oceanic	15.5	1114
Copenhagen	Cfb	Oceanic	8.9	728
Hamburg	Cfb	Oceanic	9.8	804
Nairobi	Cfb	Oceanic	18.8	674
New York	Cfa	Humid subtropical	11.9	1139
São Paulo	Cfa	Humid subtropical	19.5	1356
Sydney	Cfa	Humid subtropical	18.0	912
Beijing	Dwa	Humid continental	12.7	566
Moscow	Dfb	Warm humid continental	5.7	678
Pyongyang	Dwa	Humid continental	10.6	1043

According to *The New York Times,* the climate of New York has recently been reclassified as Cfa as the temperatures have been rising with an urban heat island effect. "New York City, after years of being considered a humid continental climate, now sits within the humid subtropical climate zone" (Collins, 2020). As might be expected the majority of the cities are located in the C or temperate zone, since these climates are very comfortable for people, with four examples in the A (equatorial) zone, and three each in the B (arid) and D (snow) zones. As noted above, climate probably has most influence on energy use, though it might also be reflected in the form of the built environment, and to an extent on the food eaten through the relationship between climate and traditional cuisines.

1.5 Governance

The type of national government can influence local or city government. Some governments are more centralised, whereas others hand out money to local authorities to deal with local issues, although money can also be raised through local taxes. In their turn local governments are often responsible for the city budget and providing essential services for the residents, such as education and some utilities. The system of central and local government is often influenced by colonisation. The British Empire left its mark through the adoption of a Westminster system of government in many countries, including Australia and New Zealand (Corbett and Veenendaal, 2016). As a starting point, Table 1.6 sets out the cities organised by continent, and their types of national and local governments.

TABLE 1.6 Summary of current government of country and city

City	*National government*	*Local government*
Cairo	Democratic republic—current president Abdel Fattah el-Sisi	Governor and Popular Assembly with elected and appointed members
Nairobi	Presidential republic—current president Uhuru Kenyatta	Nairobi City Council, with executive led by Governor. County Assembly has elected and nominated members
Ahmedabad	Democratic republic with parliamentary system of government based on the UK model—current prime minister Narendra Modi	Ahmedabad Municipal Corporation (AMC) headed by Municipal Commissioner. The corporation is elected
Beijing	Unitary Marxist–Leninist one-party socialist republic—current chairman Xi Jinping	People's Government of Beijing headed by the Mayor of Beijing.
Dubai	Federation of seven hereditary Sheikdoms—current president Khalifa bin Zayed Al Nahyan	Dubai Municipality is chaired by Deputy Ruler of Dubai
Jakarta	Presidential democracy—current president Joko Widodo	Municipality of Central Jakarta is headed by a Mayor appointed by the governor
Kuala Lumpur	Democratic system based on Westminster with an elected constitutional monarch—current prime minister (appointed by monarch) Muhyiddin Yassin	Kuala Lumpur City Hall headed by Mayor appointed by Federal Territories Minister
Pyongyang	Stalinist dictatorship—headed by Supreme Leader Marshal of the Republic Kim Jong-un	City People's Committee (administrative) and City People's Assembly (legislative)
Singapore	Parliamentary representative democratic republic—current president Halimah Yacob	Singapore is a city-state
Amsterdam	Parliamentary democracy and constitutional monarchy—current prime minister Mark Rutte	Elected Municipal (City) Council headed by a Mayor
Copenhagen	Parliamentary democracy and constitutional monarchy—current prime minister Mette Frederiksen	Elected City Council headed by a Lord Mayor
Hamburg	Democratic federal parliamentary republic—current chancellor Olaf Scholz	City state within federal Germany headed by First Mayor and President of the Senate
Moscow	Democratic federation—president Vladimir Putin	Federal city headed by a Mayor

TABLE 1.6 Cont.

City	*National government*	*Local government*
Auckland	Parliamentary democracy—current prime minister Jacinda Ardern	Auckland City Council elected local government headed by a Mayor
Sydney	Parliamentary federal democracy—current prime minister Scott Morrison	City of Sydney elected local government headed by a Lord Mayor
Havana	Marxist–Leninist socialist state—current president Miguel Díaz-Canel	City council headed by a Mayor but with a dependency on central government
New York	Democratic federal republic—current president Joe Biden	Elected city council headed by a Mayor
Buenos Aires	Presidential democratic republic—current president Alberto Fernández	Elected autonomous government headed by Chief of Government
São Paulo	Federal democratic republic—current president Jair Bolsonaro	Elected municipal council headed by a Mayor

Table 1.6 is only provided as a quick reference as systems of governance are not only complicated but also change. As an example, in 2000 Indonesia instituted a change from a centralised system to one that "...gave greater authority, political power, and financial resources directly to regencies and municipalities, bypassing the provinces" (Nasution, 2016:3). The politics of São Paulo have also changed. Neo-liberalism and its effect on economies was taken up in the United States and Western Europe in the 1980s and arrived somewhat later in Brazil (Marques, 2012).

Table 1.6 is at best a snapshot of the best available data when this chapter was written. The reason for looking at governance is to understand what lies behind the data in Table 1.7, which sets out the "democracy index" for the nations wherein the 19 cities sit. This index is compiled by the UK-based Economist Group. The index is calculated from the answers to some 60 questions that fall into the five areas of electoral process and pluralism, civil liberties, functioning of government, political participation and political culture. The aim is to measure how democratic a country is. Because COVID-19 has had an effect on democracy with a loss of civil liberties because of the need for lockdowns (The Economist Intelligence Unit, 2021:3) values in Table 1.7 are given for both 2019 and 2020.

The effect of the pandemic has been to reduce values by a small amount in established democracies but also raise them in three Asian countries (Malaysia, Singapore and China). Of the full democracies in Table 1.6, New Zealand only dropped by one point in its prompt handling of the pandemic. In the Lowy Institute ranking for performance in dealing with the pandemic, New Zealand

TABLE 1.7 Cities and national democratic indices

City	*Country*	*Type*	*Index 2019*[1]	*Index 2020*[2]
Auckland	New Zealand	Full democracy	9.26	9.25
Copenhagen	Denmark	Full democracy	9.22	9.15
Sydney	Australia	Full democracy	9.09	8.96
Amsterdam	Netherlands	Full democracy	9.01	8.96
Hamburg	Germany	Full democracy	8.68	8.67
New York	United States	Flawed democracy	7.96	7.92
Kuala Lumpur	Malaysia	Flawed democracy	7.16	7.19
Buenos Aires	Argentina	Flawed democracy	7.02	6.95
Ahmadabad	India	Flawed democracy	6.90	6.61
São Paulo	Brazil	Flawed democracy	6.86	6.92
Jakarta	Indonesia	Flawed democracy	6.48	6.30
Singapore	Singapore	Flawed democracy	6.02	6.03
Nairobi	Kenya	Hybrid regime	5.18	5.05
Moscow	Russian Federation	Authoritarian	3.11	3.31
Cairo	Egypt	Authoritarian	3.06	2.93
Havana	Cuba	Authoritarian	2.84	2.84
Dubai	UAE	Authoritarian	2.76	2.70
Beijing	China	Authoritarian	2.26	2.27
Pyongyang	North Korea	Authoritarian	1.08	1.08

1 The Economist Intelligence Unit (2020).
2 The Economist Intelligence Unit (2021).

was first of all countries at 9 January 2021, but fell to second by 13 March 2021 although achieving the same score as the first ranked country of Bhutan (Lowy Institute, 2021). Overall, the performance of small countries outranked that of larger nations. The Economist Intelligence Unit (2021:6) also noted that "...the Asia region has, so far, handled the pandemic much better than virtually any other, with lower infection and mortality rates and a fast economic rebound. By contrast, European governments were slow to act...". Of more pertinence to the discussion of democracy, the average scores showed overall authoritarian regimes scored slightly higher in the pandemic (49.6) than democratic (46.8) and hybrid (42.8) regimes, although differences are not large as noted by the Lowy Institute (2021).

> No single political system has, at this point, stood out as being significantly or consistently more effective at managing the health crisis. Countries with authoritarian models performed best at the beginning and end of the examined period. Despite a difficult start, democracies improved several months into the pandemic, before their performance eventually regressed in later stages. By contrast, many hybrid regimes...appeared least able to meet the challenge.

Despite the pandemic, the Russian Federation and Brazil also saw a rise in their albeit low democracy indices and this despite President Putin changing the constitution so that he can continue in power after his terms ends. Both Cuba and North Korea saw no change, noting that North Korea has the lowest index of all countries cited. It may come as a surprise that the United States is classed as a flawed democracy. This change from full to flawed democracy occurred in 2016 when the country "fell below the threshold for a 'full democracy' ... owing to a further decline in public trust in US institutions, a development that preceded the election of Donald Trump that year and indeed helps to explain his success in winning the presidency" (The Economist Intelligence Unit, 2020:40).

1.6 The economy and well-being

National statistics have been used to give a sense of what life on average could be like in the 19 cities based on four measures. The Human Development Index is compiled by the United Nations Development Programme (UNDP). It is calculated from average life expectancy, statistics to do with education including number of years of schooling and the standard of living as measured by per capita Gross National Income (GNI). An HDI over 7.0 is considered to be high and over 0.79 very high (Wackernagel et al., 2017). This index was developed as a better way of measuring standard of living than Gross Domestic Product (GDP), which is a measure of the national economic activity level. GDP measures everything that is bought and sold, so by definition it excludes voluntary work. This means if more women enter the labour force and the equivalent number of men do not leave it to care for the home, then childcare, house cleaning and gardening may well be purchased rather than being unpaid labour. This monetary exchange will increase GDP, but nothing new is happening in terms of living standards than was happening before. Neither index is perfect since average per capita income ignores inequality, whether this is between rich and poor in a society or gender inequality when it comes to level of pay for the same work. The Gini coefficient, named after its Italian developer Corrado Gini (Ceriani and Verme, 2012), is a measure of income inequality in a society, where "any value below 0.2 represents near perfect income equality, 0.2–0.3 shows relative equality, 0.3–04 represents adequate equality, 0.4–0.5 suggests a large income gap, and a value above 0.5 represents a severe income gap" (Garcia et al., 2021:156). Table 1.8 shows values for these indicators for the countries in which the 19 cities are located, with values taken from the same source wherever possible for consistency in their calculation.

Only Nairobi and Ahmadabad have an HDI that is not considered high, with Havana, São Paulo, Jakarta and Cairo having a high HDI and all other cities a very high value. There is a weak negative correlation for this very small sample of cities between HDI and the Gini coefficient ($r = -0.4$), suggesting that as HDI increases inequality drops. This same relationship has been noted by others. In their larger sample of 32 countries with mixed levels of development, Grimm et al. (2010) found "inequality in human development seems to be clearly negatively related to

TABLE 1.8 Cities, nations, HDI, GDP (US dollars), GNI (US dollars) and Gini coefficient ranked by HDI for 2019 where possible

City	*Nation*	*HDI*[1]	*GDP*[2]	*GNI*[1]	*Gini coefficient*[4]
Hamburg	Germany	0.947	46,232	55,314	0.32 (2016)
Sydney	Australia	0.944	54,763	48,085	0.34 (2014)
Amsterdam	Netherlands	0.944	53,053	57,707	0.28 (2018)
Copenhagen	Denmark	0.940	60,657	58,662	0.28 (2018)
Singapore	Singapore	0.938	64,103	88,155	0.38 (2019)[5]
Auckland	New Zealand	0.931	43,229	40,799	0.33 (2018)[6]
New York	United States	0.926	65,134	63,826	0.41 (2018)
Dubai	UAE	0.890	43,103	67,462	0.26 (2018)
Buenos Aires	Argentina	0.845	10,041	21,190	0.43 (2019)
Moscow	Russian Federation	0.824	11,606	26,157	0.38 (2018)
Kuala Lumpur	Malaysia	0.810	11,414	27,534	0.41 (2015)
Havana	Cuba	0.783	9,296	8,621	N/A
São Paulo	Brazil	0.765	8,755	14,263	0.53 (2019)
Beijing	China	0.761	10,004	16,057	0.39 (2016)
Jakarta	Indonesia	0.718	4,136	11,459	0.38 (2019)
Cairo	Egypt	0.707	3,161	11,466	0.32 (2017)
Ahmedabad	India	0.645	2,116	6,681	0.36 (2011)
Nairobi	Kenya	0.601	1,817	4,244	0.41 (2015)
Pyongyang	North Korea	N/A	688[3]	N/A	N/A

1 Conceição (2020:343–345).
2 United Nations Statistics Division (2021).
3 UNData (2021) (value for 2020 as 2019 not available).
4 World Bank Group (2021) (the year of the data is given in brackets).
5 Hirschmann (2021).
6 knoema (n.d.).

the average level of human development." They also noted that the poorest quintile in rich countries fared little better than the richest in poor countries. The low Gini for Denmark (0.28) is not unexpected since Denmark is famed for being the society in Europe that pays the most taxes to support a welfare system (OECD, n.d.), and in his extensive investigation of the relationship between inequality and taxation, to reduce inequality Picketty (2017:665) advised "a progressive annual tax on individual wealth—that is, on the net value of assets each person controls." This goes against the neoliberalist idea of trickle-down economics, whereby cutting taxes on the rich and businesses would spur economic growth and thus benefit the poor through job creation, to replace it with the economic sponge effect, whereby the rich suck up more than their fair share of any economic growth through osmosis.

Looking at the other values in Table 1.8, there is a stronger positive correlation between HDI and GNI (r = 0.86), suggesting that as HDI increases so does per capita gross national income, and even stronger between HDI and GDP (r = 0.89) though again this sample is very small. These latter results are hardly surprising since governments pursue a healthy GDP as a means of ensuring income well-being in society so that the apparently beneficent government will be re-elected next time. What this may be doing to inequality in society is generally ignored. However, all the values of GDP per capita are national values, and the reason people go to cities is for the economic opportunities they offer. Yeung (2011) noted "In most Asian countries, a majority of their GDP has been generated by their cities." Table 1.9 compares the national GDP per capita with that of six of the cities for which data could be found for the year 2008.

TABLE 1.9 GDP per capita for selected cities for 2008 (US dollars)

City	*National GDP per capita*[1]	*City GDP per capita*[2]
New York	48,283	73,300
São Paulo	8,878	20,300
Buenos Aires	9,147	28,000
Moscow	12,464	30,700
Singapore	40,009	47,900
Sydney	49,149	48,900

1 International Monetary Fund (2021).
2 PricewaterhouseCoopers (2009:22).

Although this data is historic and from different sources, the trend is clear in that, with the exception of Sydney, there is more economic activity in cities, and in some cities much more. It could also be argued that a high level of HDI comes at a cost to the environment because of the need to have an ever-growing economy within a limited system (the earth) whose only input is sunlight. This is explored in the next section.

1.7 The cities and the environment

To explore the environmental impact of the cities, Table 1.10 sets out the EF of the relevant nations, ranked from highest to lowest, together with their bio-capacity or available productive land area per person and the remaining balance per person, all in global hectares (gha).

In 2017 the global average bio-capacity per person was 1.6 gha while the average EF was 2.8 gha. The only city locations with surplus bio-capacity are the two pairs in South America and Oceania and Moscow in the Russian Federation. All the selected cities in Asia, a continent viewed as rising in terms of global economic importance (Lakhwinder and Gill, 2016) are in nations with negative bio-capacity, ranging from −8.4 gha in the UAE to −1.6 gha in Malaysia. As these countries continue to develop and their cities grow, the land they need will have to come from somewhere. Of even more concern are the three nations (India, Kenya and North Korea) living below the 2017 global bio-capacity of 1.6 gha, the EF that everyone on the planet could have had at that date. None has any spare bio-capacity so the only way to raise their EFs to a fair-share level would be for other nations to reduce theirs.

TABLE 1.10 Cities and per person ecological footprint and bio-capacity (GFN, 2021b)

City	*Country*	*EF gha (2017)*	*Bio-capacity (gha) (2017)*	*Surplus or deficit (gha)*
Dubai	UAE	8.9	0.5	-8.4
New York	United States	8.1	3.4	-4.6
Sydney	Australia	7.1	12.6	+5.3
Copenhagen	Demark	7.0	4.2	-2.8
Singapore	Singapore	5.9	0.1	-5.8
Moscow	Russian Federation	5.5	7.0	+1.5
Amsterdam	Netherlands	5.0	0.8	-4.2
Hamburg	Germany	4.7	1.5	-3.2
Auckland	New Zealand	4.4	9.2	+4.9
Kuala Lumpur	Malaysia	3.9	2.3	-1.6
Beijing	China	3.7	0.9	-2.8
Buenos Aires	Argentina	3.3	6.6	+3.3
São Paulo	Brazil	2.8	8.6	+5.8
Havana	Cuba	1.8	0.8	-1.0
Cairo	Egypt	1.8	0.4	-1.4
Jakarta	Indonesia	1.7	1.2	-0.5
Ahmedabad	India	1.2	0.4	-0.8
Nairobi	Kenya	1.0	0.5	-0.5
Pyongyang	North Korea	0.8	0.6	-0.2

Table 1.11 compares EF per person with HDI and per capita GNI for 2017 for the nations in which the 19 cities are located, again ranked by EF. The 2019 HDI levels have been included in brackets.

The first thing to notice from Table 1.11 is that HDI has risen for all countries from 2017 to 2019, even for those with a very high HDI of above 0.9. As might be expected there is also a strong positive correlation (r=0.81) between EF and HDI and also between EF and GNI (r=0.86).

Improved HDI and more income, both of which are currently linked to economic growth, mean a rise in EF. Given the human population is already living beyond the bio-capacity of the planet, this raises the question as to how much human development is possible. Should any increase in HDI be restricted to countries with low HDI levels, meaning that those with high HDI will have to find ways to reduce their EFs considerably and free up bio-capacity? Considerable changes are necessary, both to reduce global average EF and improve living conditions for those already living with very low environmental impact because they have so little. As living in cities tends to increase GDP, it appears that increased global urbanisation is not the best place from which to start.

TABLE 1.11 EF, HDI and GNI per capita (2017)

City	*Country*	*EF gha*	*HDI*[1] *(HDI 2019)*	*GDP per capita*[1]
Dubai	UAE	8.9	0.863 (0.890)	67,805
New York	United States	8.1	0.924 (0.926)	54,941
Sydney	Australia	7.1	0.939 (0.944)	43,560
Copenhagen	Denmark	7.0	0.929 (0.940)	47,918
Singapore	Singapore	5.9	0.932 (0.938)	82,503
Moscow	Russian Federation	5.5	0.816 (0.824)	24,233
Amsterdam	Netherlands	5.0	0.931 (0.944)	47,900
Hamburg	Germany	4.7	0.936 (0.947)	46,136
Auckland	New Zealand	4.4	0.927 (0.931)	33,970
Kuala Lumpur	Malaysia	3.9	0.802 (0.810)	26,107
Beijing	China	3.7	0.752 (0.761)	15,270
Buenos Aires	Argentina	3.3	0.825 (0.845)	18,461
São Paulo	Brazil	2.8	0.759 (0.765)	13,755
Havana	Cuba	1.8	0.777 (0.783)	7,524
Cairo	Egypt	1.8	0.696 (0.707)	10,355
Jakarta	Indonesia	1.7	0.694 (0.718)	10,846
Ahmedabad	India	1.2	0.640 (0.645)	6,353
Nairobi	Kenya	1.0	0.590 (0.601)	2,961
Pyongyang	North Korea	0.8	N/A	N/A

1 (UNDP, n.d.).

References

Alsayyad, N. (2021). *Cairo.* Retrieved from Cairo | national capital, Egypt | Britannica, 14 April 2021.

Amsterdam.info (2021). *History of Amsterdam.* Retrieved from www.amsterdam.info/basics/history/, 21 March 2021.

Ceriani, L. and Verme, P. (2012). The origins of the Gini index: extracts from Variabilità e Mutabilità (1912) by Corrado Gini. *Journal of Economic Inequality 10*(3), 421–433.

Chicca, F., Nabyl-Calliou, S., Vale, B. and Vale, R. (2018). Calculating the ecological footprints of the stories, in Chicca, F., Vale, B. and Vale, R. (eds) *Everyday Lifestyles and Sustainability: The Environmental Impact of Doing the Same Things Differently.* Abingdon: Routledge, 126–137.

Chiodelli, F. (n.d.). São Paulo, The Challenge of the Favelas. Retrieved from www.planum.net/francesco-chiodelli-sao-paulo-the-challenge-of-the-favelas, 14 April 2021.

City of New York (2021). Population—New York City Population. Retrieved from www1.nyc.gov/site/planning/planning-level/nyc-population/population-facts.page, 31 March 2021.

Climate-Data.org. (n.d.) Climate Data for Cities Worldwide. Retrieved from Climate data for cities worldwide. Climate-Data.org, 13 May 2021.

Collins, L.M. (2020). Sultry Nights and Magnolia Trees: New York City Is Now Subtropical. Retrieved from www.nytimes.com/2020/07/24/nyregion/climate-change-nyc.html, 13 May 2021.

Conceição, P. (2020). *Human Development Report 2020.* New York: United Nations Development Programme (UNDP).

Copenhagen.com (2020). Copenhagen History. Retrieved from www.copenhagen.com/historical-facts, 21 March 2021.

Corbett, J. and Veenendaal, W. (2016). Westminster in small states: comparing the Caribbean and pacific experience. *Contemporary Politics 22*(4), 432–449.

Demographia (2020). *Demographia World Urban Areas, 16th Annual Edition.* Retrieved from www.demographia.com/db-worldua.pdf, 29 March 2021.

The Economist Intelligence Unit (2020). *Democracy Index 2019: A Year of Democratic Setbacks and Popular Protest.* Retrieved from www.eiu.com/Handlers/WhitepaperHandler.ashx?fi=Democracy-Index-2019.pdf&mode=wp&campaignid=democracyindex2019, 28 May 2021.

The Economist Intelligence Unit (2021). Democracy Index 2020: In sickness and in health? Retrieved from democracy-index-2020.pdf (eiu.com), 28 May 2021.

Encylopaedia Britannica (2021). *Britannica.* Retrieved from www.britannica.com/, 29 March 2021.

Ferguson, J.L. (2021). New York History Timeline. Retrieved from www.ereferencedesk.com/resources/state-history-timeline/new-york.html, 21 March 2021.

Forster, E.M. (1910, 1941 edition). *Howards End.* Harmondsworth: Penguin.

Gracia, E., Vale, B. and Vale, R. (2021). *Collapsing Gracefully: Making a Built Environment that is Fit for the Future.* Cham: Springer.

Gillick, J.M. (2018). *A History of Kuala Lumpur, 1857–1939.* Kuala Lumpur: Malaysian Branch Royal Asiatic Society.

Global Footprint Network (GFN). (2021a). Earth Overshoot Day. Retrieved from www.overshootday.org/, 29 March 2021.

Global Footprint Network (GFN). (2021b). Map. Retrieved from https://data.footprintnetwork.org/#/exploreData, 1 June 2021.

Goudet, S.M., Kimani-Murage, E.W., Wekesah, F., Waniohi, M., and Griffiths, P.L. (2017). How does poverty affect children's nutritional status in Nairobi slums? A qualitative study of the root causes of undernutrition. *Public Health Nutrition 20*(4). 608–619.

Grimm, M., Harttgen, K., Klasen, S., Misselhorn, M., Munzi, T. and Smeeding, T. (2010). Inequality in human development: An empirical assessment of 32 countries. *Social Indicators Research 97*(2), 191–211.

Hamburg.com (n.d.). About Hamburg: History. Retrieved from www.hamburg.com/residents/about/11853158/history/, 21 March 2021.

Hirschmann, R. (2021). Gini coefficient after taxes Singapore 2011–2020. Retrieved from www.statista.com/statistics/951976/singapore-gini-coefficient-after-tax/, 31 May 2021.

Hubacek, K., Guan, D., Barrett, J. and Wiedmann, T. (2009). Environmental implications of urbanization and lifestyle change in China: Ecological and Water Footprints. *Journal of Cleaner Production 17*(14), 1241–1248.

Hugo, G., Champion, A. and Lattes, A. (2003). Toward a new conceptualization of data for demography. *Population and Development Review 29*(2), 277–297.

International Monetary Fund (IMF) (2021). World Economic Outlook Database. Retrieved from www.imf.org/en/Publications/WEO/weo-database/2021/April/weo-report, 1 June 2021.

knoema (n.d). New Zealand—GINI index. Retrieved from https://knoema.com/atlas/New-Zealand/topics/Poverty/Income-Inequality/GINI-index, 31 May 2021.

Kottek, M., Grieser, J., Beck, C., Rudolf, B. and Rubel, F. (2006). World Map of the Köppen-Geiger climate classification updated. *Meteorologische Zeitschrift 15*(3), 259–263.

Lakhwinder, S. and Gill, A. (2016). Emergence of innovative manufacturing firms across Asia. *Seoul Journal of Economics 29*(1), 113–149.

Lowy Institute (2021). Covid Performance Index. Retrieved from https://interactives.lowyinstitute.org/features/covid-performance/, 31 May 2021.

Lui, L. and Lei, Y. (2018). An accurate ecological footprint analysis and prediction for Beijing based on SVM model. *Ecological Informatics 44,* 33–42.

Marques, E. (2012). Patterns of governance and government in Sao Paolo, Brazil. Conference paper presented at Second ISA Forum of Sociology, August 1–4, 2012, Buenos Aires, Argentina.

Montclos, M-A.P. de and Kagwanja, P.M. (2000). Refugee camps or cities? The socio-economic dynamics of the Dadaab and Kakuma camps in Northern Kenya. *Journal of Refugee Studies 13*(2), 205–222.

National Geographic (2021). Köppen Climate Classification System. Retrieved from www.nationalgeographic.org/encyclopedia/koppen-climate-classification-system/, 13 May 2021.

Nasution, A. (2016). Government Decentralization Programme in Indonesia. Asian Development Bank Institute. Retrieved from www.adb.org/sites/default/files/publication/201116/adbi-wp601.pdf, 17 May 2021.

OECD (n.d.). Denmark, Sweden still the highest taxed OECD economies. Retrieved from www.oecd.org/general/denmarkswedenstillthehighest-taxoecdcountries.htm#:~:text=Denmark%20had%20the%20highest%20tax,%25%2C%20against%20Sweden's%2049.5%25, 1 June 2021.

Piketty, T. (2017). *Capital in the Twenty-First Century*. Cambridge, MA: Harvard University Press.

PricewaterhouseCoopers (2009). Which are the largest city economies in the world and how might this change by 2025? Retrieved from https://pwc.blogs.com/files/global-city-gdp-rankings-2008-2025.pdf, 1 June 2021.

Rozenfeld, H.D., Rybski, D., Gabaix, X. and Makse, H.A. (2011). The area and population of cities: New insights from a *different* perspective on cities. *American Economic Review 101.* 2205–2225.

TravelChinaGuide (2021). Beijing History. Retrieved from www.travelchinaguide.com/cityguides/beijing/history.htm, 21 March 2021.

UNData (2021). Democratic People's Republic of Korea: General Information. Retrieved from https://data.un.org/en/iso/kp.html, 31 May 2021.

United Nations (UN) Department of Economic and Social Affairs (UNESCO) (2018). News. Retrieved from www.un.org/development/desa/en/news/population/2018-revision-of-world-urbanization-prospects.html, 29 March 2021.

UNESCO (2021). The Old Town of Jakarta. Retrieved from https://whc.unesco.org/en/tentativelists/6010/, 21 March 2021.

United Nations Development Programme (UNDP) (n.d.). Human Development Index and Its Components. Retrieved from hdi_table.pdf (undp.org), 2 June 2021.

UNHCR (n.d.). Inside the World's 10 Largest Refugee Camps. Retrieved from www.arcgis.com/apps/MapJournal/index.html?appid=8ff1d1534e8c41adb5c04ab435b7974b, 29 March 2021.

United Nations Statistics Division (2021). Per capita GDP at current prices—US dollars. Retrieved from http://data.un.org/Data.aspx?q=GdP&d=SNAAMA&f=grID%3A101%3BcurrID%3AUSD%3BpcFlag%3A1, 31 May 2021.

The United States Census Bureau (n.d.) QuickFacts: New York city, New York; New York. Retrieved from U.S. Census Bureau QuickFacts: New York, 31 March 2021.

VisitSingapore.com (n.d.). About Singapore. Retrieved from www.visitsingapore.com/en_au/travel-guide-tips/about-singapore/, 21 March 2021.

Wackernagel, M., Hanscom, L. and Lin, D. (2017). Making the sustainability development goals consistent with sustainability. *Frontiers in Energy Research 5*, 5pp.

Wilson, J. (2016). Nation and government—From colony to nation. *Te Ara—the Encyclopedia of New Zealand*. Retrieved from https://teara.govt.nz/en/nation-and-government/page-2, 21 March 2021.

Winarso, H. (2011). Urban dualism in the Jakarta Metropolitan Area, in Sorensen, A. and Okata, J. (eds) *Megacities: Urban Form, Governance, and Sustainability.* New York: Springer, 163–191.

The World Bank Group (2019). Rural population (% of total population). Retrieved from https://data.worldbank.org/indicator/SP.RUR.TOTL.ZS, 29 March 2021.

The World Bank Group (2021). Gini index (World Bank estimate). Retrieved from https://data.worldbank.org/indicator/SI.POV.GINI, 31 May 2021.

World's Capital Cities (2021). Capital Facts for Jakarta, Indonesia. Retrieved from www.worldscapitalcities.com/capital-facts-for-jakarta-indonesia/, 31 March 2021.

World Population (2020). Moscow Population. Retrieved from www.populationu.com/cities/moscow-population, 31 March 2021.

World Population Review (2021). World City Populations 2021. Retrieved from https://worldpopulationreview.com/world-cities, 29 March 2021.

Yeung, Y. (2011). Rethinking Asian cities and urbanization: four transformations in four decades. *Asian Geographer 28*(1), 65–83.

2

CALCULATING THE ECOLOGICAL FOOTPRINT

2.1 Introduction

The true problem in calculating EFs is that they are always historic and in almost all cases the current situation will be having more impact on the planet than as indicated by the EF. The numbers in this research suggest that most of the global population lives under a system that has been designed to be unsustainable, and that will only worsen with urbanisation. The city-related issues that we deal with in the next five chapters (food, energy, transport, consumer goods and the built environment) cover the largest contributors to environmental degradation. There may be some double counting, but the aim here is not the calculation of absolute EFs but a calculation of those that can be compared across cities wherever possible. A brief account of the EF calculation method is given below and more detailed information can be found in Chicca et al. (2018:126–128).

2.2 EF calculation method

The method used here is based on the land needed to grow food, provide resources and sequester CO_2 emissions related with the consumption of resources. Doing this is more complicated than it seems. Land in each country has a different capacity to absorb CO_2 and to produce energy and food, which are often the largest environmental impact items. For instance, a piece of land in Brazil may be very productive when compared with the same amount of land in the Sahara Desert. A wider range of food can probably be produced in Brazilian soil, CO_2 can be absorbed by original forests and energy can be produced through local crops, such as sugar cane. In the Sahara, the only option for energy may be through the use of wind and sun, or on occasion oil may be found underground. To produce food in the Sahara, a large infrastructure would need to be created to irrigate the soil. This is the reason

DOI: 10.4324/9781003045847-3

behind the lack of settlements in the true Saharan Desert. However, this difference poses a question: What hectare of land is more environmentally valuable, a hectare in Brazil or the same hectare in the Sahara? Although it is acceptable to consider both places as environmentally important, the answer to this problem needs to be developed further. To solve this, Wackernagel and Rees (1996) devised the global hectare (gha), which is a hectare of land with the global average productivity. This has been refined by the Global Footprint Network (GFN), which provides figures for bio-capacity per country in gha per person, and this forms the basis of the calculations in this book.

Although we have used the concept and data from the GFN, the calculations for the specific impact follow a similar but slightly different methodology. The bio-capacity data is used to calculate a country index for converting hectares of land in a given country into gha. The total bio-capacity of a country, indicative of the land available for the resources it consumes, is calculated following the formula below:

$$\text{Bio}_{cap} = B_v \times Pop$$

Bio_{cap} = Total bio-capacity expressed in gha
B_v = Bio-capacity value per capita from GFN
Pop = Country's population

This index allowed us to calculate gha, for instance from food, as long as we have the productivity data for that food item in that country. From the Bio_{cap}, it is easy to establish the relationship between a country's bio-capacity expressed in gha and the country's area in hectares. This mathematical relationship is the index. The index is heavily affected by the population density of the country; small nations with high populations generally have a large conversion index. The higher the index the higher the global environmental cost of living in the country. The worst potential situation would be a country of small area with a very high population occupying land that would be very efficient in absorbing CO_2.

The next step is to transform the consumption of a particular item into land area, as this gives the amount of land necessary to provide the major items for the cities. The base of our calculation of impact is the amount of land necessary to produce a specific product and the amount of land necessary to absorb the emissions for that product's production, transportation, consumption and disposal. The index is applied to the data in hectares to give the impact in terms of EF. We are aware that many, and in some cases, most of the products to supply the cities often come from beyond national borders. Despite this, we have used the land productivity from the countries in which each city is located. We believe that as populations should be entitled to use the natural resources from their own countries, their environmental impact and the capacity of their land to absorb CO_2 should also be measured in proportion to their consumption. If a certain population lives in a country with low forest areas, poor soil and lack of water, they would need to be extra careful if their EF were large. Lacking their own resources, they will necessarily depend on someone else's land to provide what they need and cover their environmental impact. We have also produced a calculation for CO_2 emissions, for which the methodology is presented below. How other calculations were done for each issue are described in the relevant chapters.

2.3 CO_2 emissions calculations

The calculations for the CO_2 emissions (or CO_2 equivalents—CO_{2eqv}) were carried out using referenced research from previous investigations or reliable international institutions. How emissions were calculated is also described in the relevant chapters. These emissions then needed to be converted to land using the sequestration capacity according to the average for each climate determined by the IPCC (2007) as given in Table 2.1.

The emissions calculation uses the following information:

- Climate Sequestration Capacity Index in tonnes/ha/year → c_i
- Emissions in kg → f_e
- Land for CO_2 sequestration in ha→L

The equation is

$$L = \frac{f_e}{c_i}$$

The result in hectares is then multiplied by the country index C_i and divided by the number of people.

$$EF = \frac{L \star C_i}{n}$$

TABLE 2.1 Carbon dioxide sequestration rates for different climates

Climate	*Capacity for CO_2 sequestration in tonnes/ha year*
Tropical	6.0
Moderate	3.0
Boreal	0.8

References

Chicca, F., Vale, B. and Vale, R. (2018). *Everyday Lifestyles and Sustainability—The Environmental Impact of Doing the Same Things Differently*. London: Routledge.

Global Footprint Network (2017). Data Global Footprint Network. Retrieved from http://data.footprintnetwork.org/, 18 August 2021.

IPCC (2017). Land use, land change, and forestry. Retrieved from www.ipcc.ch/ipccreports/sres/land_use/index.php?idp=151, 23 August 2021.

Wackernagel, M. and Rees, W. (1996). *Our Ecological Footprint—Reducing Human Impact on the Earth*. Gabriola Island: New Society Publisher.

3
FOOD

3.1 Problems with food data

Although one of the most relevant environmental impacts, it is hard to measure the impact of food precisely, particularly in large-scale regions or cities. Bouwman (1997), in the United Nations Food and Agriculture Organization (FAO) 1997 report *Long-Term Scenarios of Livestock–Crop–Land Use Interactions in Developing Countries*, highlighted that not all environmental consequences from food production can be quantitatively evaluated. Although there has been considerable development in data collection and analysis, the challenges around the measurement of the environmental impact of food at the large scale have not been solved.

The environmental impact of any food item is intrinsically complicated. The unpredictability and dependence on weather and soil add an extra layer of complexity. On many occasions, researchers will come across very different numbers for CO_{2eqv} emissions for the same agricultural product. For red meat, it is possible to find figures from 36 kg CO_{2eqv}/kg to as low as 7.5 kg of CO_{2eqv}/kg (Lynch and Pierrehumbert, 2019). There are many reasons for the differences found for the same food products. Although some of these may have come from sketchy or biased research, the differences could simply have occurred because of different approaches. The calculation of emissions from the entire life cycle analysis of food is long and complex with a huge number of variables, one being fertilisers. According to the IPCC (Intergovernmental Panel on Climate Change) (2014) agricultural products from cropland and pastures represent around 11% of greenhouse gas (GHG) emissions and around 30–40% of these emissions are related to fertilisers. In some specific circumstances, such as winter wheat, the impact of fertilisers can be as high as 80% of the total product emissions (Brentrup et al., 2016). The impact of fertilisers cannot be understated. Fertilisers are associated with N_2O (nitrous

DOI: 10.4324/9781003045847-4

oxide), which is a potent greenhouse gas, one kilogram of N_2O being equal to 298 kilograms of CO_2 (Environmental Protection Agency, 2021). In 2019 in the United States, N_2O represented 7% of all GHG emissions (Environmental Protection Agency (EPA), (2021), with 75% of N_2O emissions being directly related to agricultural soil management.

To determine the precise impact of any agricultural product, it is necessary to identify how that product is grown; whether the product comes from an organic farm; what the weather was like during the crop cycle as the weather may determine irrigation or further soil treatment; the type and amount of fertiliser used; if a greenhouse was used; and how much mechanisation the farm relies on, that is whether input comes from tractors or people. All the potential impacts listed above happen before the crop leaves the farm. The way that the food item travels from the farm to our plates, and how we prepare the food and dispose of any leftovers will dramatically change the impact of food.

Not all the differences in CO_2 emissions for food are the outcomes of different scientific approaches. The food industry is known for its strong lobby. The efforts of the industry's fight against climate change policies have been well documented. For example, Brulle (2014) identified the financial effort made by the food industry in the United States to maintain the status quo, calling it the Climate Change Counter-Movement (CCCM). The CCCM is not new and has been acting against climate change policies since 1989 just after the creation of the Intergovernmental Panel on Climate Change (Antonio and Brulle, 2011). Lazarus et al. (2021) identified the attempts of dairy and meat producers to undermine political efforts to mitigate climate change and their efforts to hire their own academics to minimise the link between animal agriculture and GHG emissions. Dunlap and McCright (2010) revealed the mechanism by which corporations intentionally fund "junk science" aiming to de-legitimise public opinion and policy makers. These deliberate actions to produce weak data form a major and complex problem for measuring the impact of food.

Measuring the consumption of food in large scale areas is also complicated. Urban areas often do not have clear boundaries, making it hard to identify the exact population and their food suppliers. Frequently where food comes from is also an issue, as food from sources such as private gardens, community gardens, small local farms and allotments may not be included in accounts.

Another factor that may affect the data is related to waste. Food loss or waste represents 33% of total food produced (FAO, 2011; FAO, 2015). In some cases, it can be as high as 40% (Porter et al., 2016). When considering home-made food, the waste can be included in the household total, however, processed and take-away food is more complicated. It is likely that part of the waste is not accounted for in the environmental impact of such food, despite the relevance of the number.

The hardest part of measuring the environmental impact of food is related to the data itself; for this book, this means finding it and ensuring that the data is comparable between all the cities. As it is virtually impossible to measure all food consumption in large urban areas precisely, all such research must rely on third party data. In addition, there are specific problems related to some of our sample cities. For example, Pyongyang, Havana, Nairobi and Ahmedabad do not provide many options for accessing data. Several cities are regional or international hubs, and the food entering them may not be consumed by their inhabitants, but processed, stored or further distributed to other locations.

The conclusion is, even if all the data were available, measuring and determining the impact from food consumption in urban areas will be very hard. A specific comparative criterion is needed. The goal of undertaking such measurement goes beyond the precision of the final number. It aims to provide a meaningful comparison between urban areas and tries to identify why certain regions have a massive difference in the impact of their food consumption. The comparison is more important than the cold precision of the environmental impact number. The idea is to come up with a method that allows for comparison between cities and support for an understanding of why the official data of these cities may not be useful for comparisons.

3.2 How to measure the environmental impact of the food that really matters

Despite all the challenges with data, there are ways to provide an educated assessment of the environmental impact of food in urban areas. Measuring the impact could start with local sample data, which can then be extrapolated to the entire urban area. Some local data could be combined with state or even national data. Baabou et al. (2017) have compared the environmental impact of 19 Mediterranean cities using the ecological footprint (EF). In their research, they presented a comprehensive literature review of how other researchers have dealt with the data issue. They showed that both top-down and bottom-up approaches have been used for different cities. In most cases, the top-down approach was applied. This means using national data as the primary data and adapting it to the local case. The cases that used a bottom-up approach only used local data to calculate the environmental impact. This may work for one city, but in order to compare different areas, the same criteria need to be used for all the urban areas involved. In their comparisons, Baabou et al. (2017) used a monetary multi-regional input–output (MRIO) approach, breaking down the impact by consumption categories. For the impact of food, in this book we have used national data and, where possible, have tried to find data for the individual city that could be used to adjust the national data. Decoding the reasons that push cities to become more sustainable or more unsustainable is our primary interest. A top-down approach was used, which means that we started from the national data, for which equalisation could be ensured, since this data was most likely to be both reliable and available.

The environmental impact of food was calculated using CO_{2eq} and the ecological footprint. Besides the different approaches that generate a different impact for the same food, there is also a massive difference between food items when it comes to emissions, as shown in Table 3.1. The first column shows the emission factor from Carlson-Kanyama and Gonzálvez (2009) and the second column emission factors from different sources.

TABLE 3.1 CO_2 emissions for selected foods

Commonly consumed foods	*kg CO_2 equivalents/kg product*	
Carrot	0.42	0.14 (Raghu, 2015)
Potatoes (cooked)	0.45	0.15 (Moudrý et al., 2013)
Rice (cooked)	1.3	2.12 (Xu et al., 2013)
Green Beans	1.3	0.14 (Healabel, 2021a)
Pork	9.3	5.16 (FAO, 2021a)
Cheese	11.0	14.0 (FAO, 2021a)
Beef (cooked)	30.0	21.1 (FAO, 2021a)

Regardless of the considerable difference between the columns dependent on the source of the calculations, there is a common issue in both, this being the difference in CO_2 emissions between vegetables and livestock products. Livestock products have at least ten times the impact of plant-based products. This remarkable difference is a common finding in all research that we assessed. The United Nations Food and Agriculture Organization (FAO) has released a comprehensive report clarifying the impact of animal products. In *Livestock's Long Shadow*, the FAO (2006:xxi) has indicated the scale of the impact from livestock-based food: "The livestock sector is a major player, responsible for 18 percent of greenhouse gas emissions measured in CO_2 equivalent. This is a higher share than transport."

Chicca et al. (2018) have suggested another explanation for why livestock products have a much larger impact. Livestock requires more land to produce the same amount of food energy. Tables 3.2 and 3.3 show how much energy could be produced by comparing different food types with one unit of energy from meat and from milk. In both tables, the yield data were extracted from FAO (2006) and the energy from USDA (2021).

TABLE 3.2 The energy generated from the same amount of land in different countries compared with one unit of energy from meat

Country	*Meat*	*Potatoes*	*Wheat*
New Zealand	1	160.4	122.6
Argentina	1	112.9	44.1
Brazil	1	46.3	15.8
United States	1	40.3	10.8

TABLE 3.3 The energy generated from the same amount of land in different countries compared with one unit of energy from milk

Country	*Milk*	*Potatoes*	*Wheat*
New Zealand	1	22.80	17.49
Argentina	1	285.36	111.57
Brazil	1	95.07	32.39
United States	1	52.71	14.16

Meat and milk also have a vastly unfavourable energy input–output ratio, and more energy (largely in the form of fossil fuels) is necessary to produce the meat and milk than the food energy provided when eaten. Pimentel and Pimentel (2003) suggested that meat requires 40 times more energy to produce than the energy provided, while milk requires 13 times the energy. For potatoes, on the other hand, for each unit of energy used, 2.76 units of energy are generated (Pimentel, 2009).

The impacts of dairy products and meat are much greater than most other types of food, but any diet has some environmental impact. If part of this diet includes meat and dairy, probably the majority of its impact will occur because of their consumption. The amount of GHG emissions, the land area necessary to produce meat and dairy compared with other kinds of crops and the poor energy ratio have led us to pay close attention to the consumption of meat and dairy products. The environmental impact of food in our calculations is based on the consumption of meat and dairy and the rest of the diet is considered to be supplied by unprocessed food to achieve the minimum number of calories indicated by the United Nations. This will probably give an underestimate of the overall environmental impact.

3.3 Methodology and results

The calculation of the CO_{2eqv} combines the emissions from meat and dairy products and the amount consumed by each city. The first step in calculating the CO_{2eqv} for food was to assume that individuals in all 19 cities consume 2,500 kcal per day as recommended by FAO (FAO, 2021c). FAO provides the data for the consumption of meat, milk and cheese in all countries; therefore, the calories from meat and dairy can be calculated. To achieve the minimum intake required by the United Nations, we have added to the impact of meat and dairy, the calories necessary to achieve 2,500 kcal. The additional kcals are from three different groups of food, as recommended by the British National Health Service (NHS) (NHS, 2021), as seen in Table 3.4. To reach the minimum required, the calories are added following the recommendation from NHS of 40%, 40% and 20% of calories coming from groups 1, 2 and 3, respectively. For example, if an individual consumes 1,500 kcal from meat and dairy meaning 1,000 kcal are lacking, 400 kcal will come from each of groups 1 and 2, and the remaining 200 kcal will be from group 3. In Table 3.4, the portion sizes and kcal are from Wolfram|Alpha (2021), while references for emissions are given in the right-hand column.

TABLE 3.4 Additional food items by group

		Portion in g	*kcal per portion*	*CO_{2eqv}kg/kg of product*	
Group 1	Cabbage	70	15	0.12	HEALabel (2021b)
	Banana	20	35	0.48	Berners-Lee (2020)
	Carrot	120	30	0.38	Carlsson-Kanyama and González (2009)
	Pea	110	80	0.49	Nette et al. (2016)
	Apple	93	50	0.55	Berners-Lee (2020)
	Broccoli	89	30	0.30	Eriksson et al. (2021)
Group 2	Potato	110	140	0.15	Moudrý et al. (2013)
	Rice[a]	63	200	2.12	Xu et al. (2013)
	Breads	30	80	1.39	Espinoza et al. (2011)
Group 3	Lentils	35	100	0.90	Environmental Working Group (2011)
	Egg	50	70	4.80	FoodFootprint (2021)
	Beans	130	100	1.50	Harwatt et al. (2017)

a The environmental impact from rice is an average from different crops (Xu et al., 2013).

Table 3.5 shows the consumption of meat for each of the cities (FAO, 2021c). The cities are ranked according to the consumption of red meat (cattle). Terms in brackets are as used in the formula for emissions from meat.

TABLE 3.5 Meat consumption kg/capita/year

Cities	*Red meat (cattle)*	*Poultry (chicken)*	*Pigs (pork)*	*Mutton and goat (sheep)*
Buenos Aires	55.48	38.66	10.51	1.56
São Paulo	39.25	45.00	12.60	0.62
New York	36.24	50.01	27.64	0.43
Sydney	33.86	46.12	24.14	9.87
Copenhagen	28.46	26.75	24.87	0.92
Auckland	22.49	34.98	22.29	18.91
Amsterdam	17.67	23.9	36.36	0.94
Moscow	16.98	27.5	25.66	1.40
Hamburg	13.16	17.75	51.81	0.73
Cairo	12.88	14.43	0.02	1.58
Nairobi	9.54	0.47	0.29	1.86
Havana	7.27	2.55	20.40	0.85
Kuala Lumpur	6.52	41.01	7.76	0.94
Dubai	5.64	41.04	0	9.08
Beijing	5.23	13.73	38.6	3.09
Singapore	3.00	34.00	22.00	2.00
Jakarta	2.59	7.50	2.98	0.46
Pyongyang	0.88	1.76	4.70	0.63
Ahmedabad	0.81	1.88	0.28	0.58

The consumption of dairy products is also responsible for a large part of the environmental impact from food items, even though it is not as bad as red meat. The cities in Table 3.6 are organised according to the consumption of milk from FAO (2021c). There is no data for Havana.

TABLE 3.6 Consumption of milk and cheese kg/capita/year

Cities	*Milk*	*Cheese*
Amsterdam	341.47	21.6
Copenhagen	277.30	28.1
Hamburg	258.70	24.7
New York	254.69	16.7
Sydney	234.49	14.7
Buenos Aires	195.08	11.6
Moscow	163.57	5.7
São Paulo	149.28	3.8
Auckland	137.25	8.2
Dubai	132.66	6.0
Nairobi	94.86	NA
Ahmedabad	84.50	0.2
Cairo	59.46	4.2
Singapore	46.10	1.87
Beijing	33.18	0.1
Kuala Lumpur	25.28	NA
Jakarta	14.82	0.01
Pyongyang	3.79	NA

The combined consumption of meat and dairy, the total calories and the calories needed to achieve 2,500 kcal as recommended by the UN are shown in Table 3.7.

TABLE 3.7 Balance of food intake of 2,500 kcal/day for 19 cities

	Calories from meat and dairy (kcal)	*Calories from Groups 1, 2 and 3 (kcal)*
Amsterdam	1681	819
Copenhagen	1582	918
Hamburg	1566	934
New York	1563	937
Sydney	1490	1010
Buenos Aires	1299	1201
Auckland	1079	1421
São Paulo	1012	1488
Moscow	953	1547
Dubai	700	1800
Beijing	520	1980

(continued)

TABLE 3.7 Cont.

	Calories from meat and dairy (kcal)	*Calories from Groups 1, 2 and 3 (kcal)*
Singapore	504	1996
Cairo	369	2131
Kuala Lumpur	359	2141
Nairobi	333	2167
Ahmedabad	245	2255
Havana	234	2266
Jakarta	117	2383
Pyongyang	66	2434
Copenhagen	1582	918
Hamburg	1566	934

The calculation of CO_{2eqv} is based on the consumption of meat and dairy as presented in Tables 3.5 and 3.6. As previously explained, the emissions from meat vary dramatically according to different sources. Poore and Nemecek (2018), for example, after assessing data from 38,000 farms, noted very different results for emissions. This variation for red meat was as high as 400% between the lowest to the highest emissions. Flysjö (2012) suggested that productivity, age of the cattle and different use of the animal (not specialised for milk or meat) lead to a different impact. The variation can be seen in Table 3.8. The CO_{2eqv} is often presented as $CO_{2eqv/kg}$ per kg of product or $CO_{2eqv/kg}$ per 100g of protein. Here, however, values have been converted to kg of product using the tables from the United States Department of Agriculture (USDA, 2021) .

TABLE 3.8 Variation in CO_{2eqv} of red meat

	CO_{2eqv} per kg of red meat	*Source*
Red Meat— Beef (cattle)	99.48	Our World in Data (2021)
	27.00	Environmental Working Group (2011)
	13.00	Grabolle and Loitz (2007)
	48.70	FAO (2021c)

After considering different sources, the data from the Global Livestock Environmental Assessment Model (GLEAM) from FAO was used for further calculation (FAO, 2021b). This data was chosen for several reasons. First, because it presents emissions for all kinds of livestock meats and milk. Second, it presents the emissions per region considering the average production for each region as well as a global average emission for each livestock product, and finally, it presents a breakdown of the emissions in comprehensive terms, including such issues as fertiliser and crop residue and manure management. The final number is presented in CO_{2eqv}. Because GLEAM presents the number per region, we could use different emissions for each region making the calculation more precise. The global emission was helpful as a complement to the calculations. Some countries are unable to produce the amount of meat required for their internal consumption. The quantities for imports were calculated looking at the new set of data from the Global Footprint Network (GFN, 2021), which includes the biocapacity, the production and the consumption. The final formula for the emissions from meat is

$$E_{\mathrm{m}} = (M \times mCO_2) + (S \times sCO_2) + (P \times pCO_2) + (C \times mCO_2)$$

where

E_{m} = Emissions for meat (all kinds) in CO_{2eqv} expressed in kg per year
M = red meat consumption from Table 3.4.

The other letters in the formula are S for sheep and mutton consumption, P for pork and C for chicken.

The calculation is slightly different if, according to the dataset from GFN, the country is unable to produce all the livestock products it consumes. In this case, the formula needs to be used twice, once using the local emissions values and then using the global average for the proportion of imports, both using the emissions data from GLEAM. For example, if the country where the city is located produces 40% of the local consumed meat, we assumed that 40% of the emissions would be calculated using the local emissions figures from GLEAM and the remaining 60% would be calculated using the global average figure.

For dairy, FAO is the source of the consumption of milk and cheese and GLEAM the emissions from milk, but not cheese. The cheese is converted into milk using the proportion 1:10 as determined by Dairy Food Safety, an independent board accountable to the Victorian Minister for Agriculture in Australia (Dairy Food Safety, 2021).

$$E_{\mathrm{of}} = E_{\mathrm{m}} + E_{\mathrm{d}} + E_{\mathrm{r}}$$

where

E_{of} = total of emissions from food in CO_{2eqv} kg/year and
E_{m}, E_{d} *and* E_{r} are, respectively, the emissions from meat, dairy and the remaining items from food groups 1, 2 and 3 to achieve 2,500 kcal.

The results are presented in Table 3.9. The remarkable difference in CO_{2eqv} emissions provided by GLEAM, which gives figures as diverse as 54.8kg/ CO_{2eqv} to 16k/ CO_{2eqv} per kilo of red meat for East Asia and Russia, respectively, affects the final result, but even the lowest figure for CO_{2eqv} of meat is still much higher than that for vegetables.

The EF calculation was very straightforward being based on the most updated data available from GFN for each country (Global Footprint Network, 2021). It included all the impact in the category of grazing and 91% of the impact under the cropland category. Cropland production includes fibre for human consumption,

TABLE 3.9 Total CO_{2eqv} emissions for local and imported meat and dairy products and other food to achieve 2,500 kcal

	CO_{2eqv} *emissions for local production (kg/year/person)*						CO_{2eqv} *emissions for imports (kg/year/person)*								
	Bovine (cattle)	*Poultry (chicken)*	*Pigs*	*Mutton and Goat (sheep)*	*Cheese*	*Milk*	*Bovine (cattle)*	*Poultry (chicken)*	*Pigs*	*Mutton and Goat (sheep)*	*Cheese*	*Milk*	*Total CO_{2eqv} emissions from livestock products*	*Total CO_{2eqv} emissions from groups 1, 2 and 3*	*Total CO_{2eqv} / kg/year/ person*
Buenos Aires	3780.68	282.06	64.27	70.05	504.37	848.21	0.00	0.00	0.00	0.00	0.00	0.00	5549.65	197.95	**5747.60**
São Paulo	2674.69	328.32	77.05	27.84	165.22	649.07	0.00	0.00	0.00	0.00	0.00	0.00	3922.20	245.30	**4167.49**
Copenhagen	8.41	2.65	1.95	0.27	7.43	7.33	1367.75	212.28	151.06	41.61	960.87	948.22	3709.82	151.30	**3861.12**
Amsterdam	28.31	12.81	15.44	1.48	30.93	48.90	795.93	177.77	206.99	39.85	692.27	1094.39	3145.07	135.05	**3280.12**
Sydney	1098.94	370.14	219.81	282.65	335.16	534.64	0.00	0.00	0.00	0.00	0.00	0.00	2841.34	166.48	**3007.81**
Hamburg	95.34	43.04	99.53	5.20	159.97	167.55	421.40	93.85	209.67	22.00	562.75	589.41	2469.71	154.01	**2623.72**
Auckland[a]	729.92	280.74	202.96	541.52	186.96	312.93	0.00	0.00	0.00	0.00	0.00	0.00	2255.04	234.30	**2489.34**
New York	865.25	242.87	112.99	16.49	380.09	579.67	0.00	0.00	0.00	0.00	0.00	0.00	2197.36	154.52	**2351.88**
Moscow	65.02	32.37	24.79	8.33	26.43	75.84	629.83	168.44	120.29	48.87	150.43	431.69	1782.33	255.04	**2037.37**
Dubai	0.61	0.79	0.00	1.24	0.49	1.09	274.25	329.52	0.00	415.55	207.59	458.98	1690.11	296.78	**1986.89**
Nairobi	645.51	4.01	2.46	113.16	0.00	514.38	0.47	0.00	0.00	0.09	0.00	0.33	1280.42	357.20	**1637.63**
Beijing	208.55	116.71	197.69	84.98	2.27	75.28	69.30	30.04	64.64	38.53	0.94	31.28	920.22	326.43	**1246.64**
Singapore	0.00	0.00	0.00	0.00	0.00	0.00	146.22	273.65	135.52	91.75	64.85	159.87	871.86	329.01	**1200.87**
Kuala Lumpur	6.64	8.90	1.68	0.66	0.00	0.00	311.88	323.93	46.91	42.32	0.00	86.04	828.97	352.95	**1181.91**
Cairo[b]	129.70	25.91	0.03	5.85	60.91	86.23	38.02	90.59	0.10	56.53	113.61	160.84	768.33	351.31	**1119.63**
Havana	395.04	14.81	99.41	30.31	0.00	62.77	72.02	4.16	25.53	7.89	0.00	0.00	711.95	373.64	**1085.59**
Ahmedabad	18.75	5.96	0.88	13.09	0.40	169.97	24.85	9.52	1.09	16.75	4.37	184.45	450.07	371.80	**821.88**
Jakarta	66.25	40.90	9.79	8.12	0.15	21.57	67.27	32.17	9.78	11.25	0.18	27.39	294.81	392.85	**687.66**
Pyongyang	25.53	8.43	17.23	17.20	0.00	7.82	17.37	5.74	11.72	11.70	0.00	5.32	128.05	401.21	**529.26**

a GFN does not give the total production of livestock for New Zealand but as, according to Landcare Research (2020). New Zealand is an exporter of livestock products, all production was assumed to be local.

b GFN (2021) does not breakdown consumption or production for Egypt so these figures are estimates.

feed for livestock, energy crops and rubber. Cassidy et al. (2013) state that crops used for industrial production are 9% of crops by mass, meaning that 91% of cropland production is used for food, either directly or as food for livestock.

$$EF_{food} = Grazing + 91\%\ Cropland$$

where

EF_{food} is the total ecological footprint of food

Grazing is the impact of livestock products consumption from the Global Footprint Network dataset expressed in global hectares (gha)

Cropland is the impact of agricultural products from the Global Footprint Network dataset expressed in global hectares (gha).

The results are presented in Table 3.10 along with the CO_{2eqv} emissions from Table 3.9.

TABLE 3.10 The EF of food for the 19 cities

City	*Environmental impact of food in gha*	*Environmental impact of food in CO_{2eqv} emissions*
Sydney	1.67	3007.8
Copenhagen	1.57	3861.1
Buenos Aires	1.54	5747.6
Amsterdam	1.50	3280.1
São Paulo	1.44	4167.5
Auckland	1.32	2489.3
Dubai	1.26	1986.9
New York	1.17	2351.9
Moscow	0.98	2037.3
Hamburg	0.87	2623.7
Kuala Lumpur	0.81	1181.9
Beijing	0.73	1246.6
Cairo	0.69	1119.6
Singapore	0.68	1200.9
Havana	0.60	1085.6
Jakarta	0.53	687.7
Nairobi	0.39	1637.6
Ahmedabad	0.31	821.9
Pyongyang	0.20	529.3

3.4 Discussion

Food is a relevant player in environmental impact, and despite some differences, both methods (CO_{2eqv} and ecological footprint) present broadly similar results. Although CO_{2eqv} emissions are included in EF, there are other factors that impact the result. Regardless of the differences, there is a strong correlation between gha and CO_{2eqv} emissions. The *r* correlation value between CO_{2eqv} and gha is very high

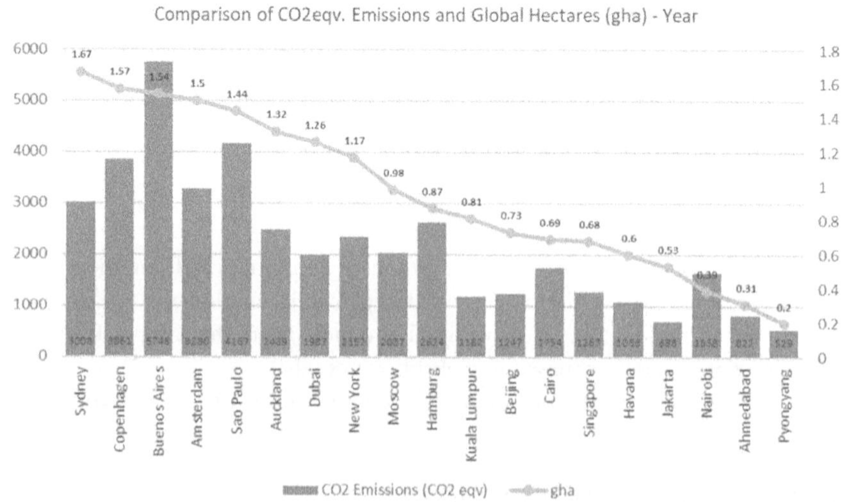

FIGURE 3.1 Comparison between CO_{2eqv} and EF in global hectares (gha)

(0.88), as seen in Figure 3.1. There are several limitations. Although we cannot precisely quantify the impact, we are able to produce a strong comparison between the cities. Some figures are worthy of comment. The Argentinian diet is high in meat consumption, being ranked first in the world for beef consumption (Euromeat, 2022), while milk consumption in Kenya is the highest in sub-Saharan Africa (3R Kenya Project, n.d.).

It is also possible to determine additional factors that affect the environmental impact of the cities, distorting our comparison. Several studies have pointed out the relationship between income and meat consumption (Goodfray et al., 2018; Mathijs, 2015; Schroeder et al., 1996; Whitnall and Pitts, 2019). Therefore, if the city has a higher average income than the rest of the country, it is likely the city will also have a higher environmental impact for food. Table 3.10 shows that the richer cities have higher food impacts. Table 3.11 shows in more detail the income per capita, for both cities and country all from the same source (Statista, 2021). There is no information for Pyongyang and Havana. It suggests that Hamburg, Amsterdam, Copenhagen, New York, Auckland, Buenos Aires, Moscow, Kuala Lumpur, São Paulo, Jakarta and Nairobi could have a larger environmental impact from food than the rest of their countries.

In addition, Chicca et al. (2018:129–130) have referred to the availability and consumption of fast food, which produces a larger environmental impact that home-made food. Mills et al. (2017) suggested that eating at home has generally more vegetables and therefore lower environmental impact and is also healthier. Religious and cultural aspects are also major players. Ahmedabad, for example, where the vast majority of the population are Hindu (The Registrar General and Census Commissioner, 2015), has very low meat consumption. Buenos Aires and São Paulo are two special cases, as the meat consumption is much higher than

TABLE 3.11 Comparison of income per capita for cities and the nations in which they sit

	GNI (City)	*GNI (Country)*
Sydney	$49,860.0	$51,019.0
Hamburg	$44,791.0	$42,065.0
Beijing	$9,689.0	$10,002.0
Singapore	$57,232.0	$57,232.0
Amsterdam	$72,567.0	$50,921.3
Copenhagen	$59,766.0	$56,916.0
New York	$62,479.0	$59,471.0
Auckland	$44,678.0	$41,662.0
Pyongyang	N/A	$684.0
Dubai	$39,832.0	$63,589.0
Buenos Aires	$13,698.0	$11,989.0
Moscow	$11,262.0	$10,754.0
Kuala Lumpur	$10,909.0	$10,127.0
Havana	N/A	$8,064.0
São Paulo	$9,991.0	$9,367.0
Cairo	$2,619.0	$2,660.0
Jakarta	$4,144.0	$4,039.0
Ahmedabad	$2,144.4	$2,631.0
Nairobi	$1,680.0	$1,660.9

countries with similar or even higher incomes. Singapore is the opposite, as despite the high average income, the consumption of meat is lower than in countries with similar incomes.

In conclusion, as with all the other areas we have studied here, it seems to be the rich that are causing all the problems.

References

Antonio, R.J. and Brulle, R.J. (2011). The Unbearable Lightness of Politics: Climate Change Denial and Political Polarization. *The Sociological Quarterly 55*(2), 195–202.

Baabou, W., Grunewald, N., Ouellet-Plamondon, C., Gressot, M. and Galli, A. (2017). The Ecological Footprint of Mediterranean cities: Awareness creation and policy implications. *Environmental Science & Policy 69*, 94–104.

Berners-Lee, M. (2020). *How Bad Are Bananas?: The Carbon Footprint of Everything.* London: Profile Books Ltd.

Bouwman, A. (1997). *Long-Term Scenarios of Livestock-Crop-Land Use Interactions in Developing Countries.* Rome: FAO United Nations.

Brentrup, F., Hoxha, A. and Christensen, B. (2016). Carbon footprint analysis of mineral fertilizer production in Europe and other world regions. The 10th International Conference on Life Cycle Assessment of Food. LCA food 2016.

Brulle, R. (2014). Institutionalizing delay: foundation funding and the creation of U.S. climate change counter-movement organizations. *Climate Change 122*(4), 681–694.

Carlsson-Kanyama, A. and González, A. (2009). Potential contributions of food consumption patterns to climate change. *The American Journal of Clinical Nutrition 89*(50), 1704–1709.

Cassidy, E.S., West, P.C., Gerber, J.S. and Foley, J.A. (2013). Redefining agricultural yields: from tonnes to people nourished per hectare. *Environmental Research Letters 8*(3), 034015.

Chicca, F., Vale, B. and Vale, R. (2018). *Everyday Lifestyle and Sustainability—The Environmental Impact of Doing the Same Things Differently*. London: Routledge.

Dairy Food Safety. (2021). Home. Retrieved from www.dairysafe.vic.gov.au/, 26 November 2021.

Dunlap, R.E. and McCright, A.M. (2010). Climate change denial: sources, actors, and strategies. In C. Lever-Tracy (ed). *Routledge Handbook of Climate Change and Society* . London: Routledge, 240–260.

Environmental Protection Agency (2021). Greenhouse Gas Emissions. Retrieved from www.epa.gov/ghgemissions/overview-greenhouse-gases, 1 November 2021.

Environmental Working Group. (2011). Lifecycle Assessments: Methodology and Results - Meat Eater's Guide to Climate Change + Health. Retrieved from https://static.ewg.org/reports/2011/meateaters/pdf/methodology_ewg_meat_eaters_guide_to_health_and_c limate_2011.pdf, 25 November 2021.

Eriksson, M., Bartek, L., Löfkvist, K., Malefors, C. and Olsson, M.E. (2021). Environmental assessment of upgrading horticultural side streams—The case of unharvested broccoli leaves. *Sustainability, 13*(10), 15pp.

Espinoza, N., Stichnothe, H. and Azapagic, A. (2011). The carbon footprint of bread. *The International Journal of Life Cycle Assessment, 16*, 351–365.

Euromeat (2022). Meat consumption drops in Argentina. Retrieved from Meat consumption drops in Argentina. Euromeatnews.com, 23 May 2022.

FAO. (2006). *Livestock's Long Shadow*. Rome: United Nations.

FAO. (2011). *Global Food Losses and Food Waste: Extent, Causes and Prevention*. Rome: United Nations.

FAO. (2012). *Livestock and Landscape*. Rome: United Nations.

FAO. (2015). *Food Wastage Footprint & Climate Change*. Rome: United Nations.

FAO. (2021a). FAOSTAT. Retrieved from FAO: www.fao.org/faostat/en/#data, 12 November 2021.

FAO. (2021b). Global Livestock Environmental Assessment Model (GLEAM). Retrieved from Global Livestock Environmental Assessment Model (GLEAM): www.fao.org/gleam/model-description/en/, 25 August 2021.

FAO. (2021c). Prospects for food and nutrition. Retrieved from www.fao.org/3/a0607e/a0607e02.pdf, 22 November 2021.

Flysjö, A. (2012). Greenhouse Gas Emissions in Milk and Dairy Product Chains. PhD Thesis. Aarhus University.

FoodFootprint. (2021). Climate impact CO_2 eggs (cooked). Retrieved from https://foodfootprint.nl/en/foodprintfinder/eggs-cooked/, 25 November 2021.

Global Footprint Network. (2021). National Footprint and Biocapacity Accounts. Oakland, California, CA: Global Footprint Network.

Godfray, H.C.J., Aveyard, P., Garnett, T., Hall, J.W., Key, T.J., Lorimer, J., Pierrehumbert, R.T., Scarborough, P., Springmann, M. and Jebb, S.A. (2018). Meat consumption, health, and the environment. *Science 361*(6399), 8.

Grabolle, A. and Loitz, T. (2007). Pendos CO_2-Zähler (1st ed., Vol. 1). Chicago: Pendo Press.

Harwatt, H., Sabaté, J., Eshel, G., Soret, S. and Ripple, W. (2017). Substituting beans for beef as a contribution toward US climate change targets. *Climatic Change, 143*(1–2), 261–270.

HeAlabel. (2021a). Green Beans. Retrieved from Healabel Health Environment Animals Laboreres: https://healabel.com/g-ingredients/green-beans, 1 November 2021.

HeALabel. (2021b). Is cabbage good or bad? Retrieved from https://healabel.com/c-ingredients/cabbage, 1 November 2021.

IPCC. (2014). *Climate Change 2014: Mitigation of Climate Change (Final draft Report of Working Group III. Contribution to the Fourth Assessment Report of the IPCC)*. Geneva: United Nations.

Landcare Research—Manaaki Whenua (2020). Manaaki Whenua—Landcare Research is the Crown Research Institute for our land. Retrieved from www.landcareresearch.co.nz/, 2 November 2021.

Lazarus, O., McDermid, S. and Jacquet, J. (2021). The climate responsibilities of industrial meat and dairy producers. *Climate Change 165*(30), 21.

Lynch, J. and Pierrehumbert, R. (2019). Climate impacts of cultured meat and beef cattle. *Frontiers in Sustainable Food Systems 3*, 11.

Mathijs, E. (2015). Exploring future patterns of meat consumption. *Meat Science 109*, 112–116.

Mills, S., Brown, H., Wrieden, W., White, M. and Adams, J. (2017). Frequency of eating home cooked meals and potential benefits for diet and health: cross-sectional analysis of a population-based cohort study. *The International Journal of Behavioral Nutrition and Physical Activity 14*(1), 11.

Moudrý, J., Jelínková, Z., Jarešová, M., Plch, R., Moudrý, J. and Konvalina, P. (2013). Assessing greenhouse gas emissions from potato production and processing in the Czech Republic. *Outlook on Agriculture 42*(3), 179–183.

NHS (National Health Service) (2021). The Eatwell Guide. Retrieved from www.nhs.uk/live-well/eat-well/the-eatwell-guide/, 23 November 2021.

Nette, A., Wolf, P., Schlüter, O. and Meyer-Aurich, A. (2016). A comparison of carbon footprint and production cost of different pasta products based on whole egg and pea flour. *Foods 5*(4), 8.

Our World in Data (2020). The carbon footprint of foods: Are differences explained by the impacts of methane? Retrieved from https://ourworldindata.org/carbon-footprint-food-methane, 1 October 2021.

Our World in Data (2021). Greenhouse gas emissions per kilogram of food product. Retrieved from https://ourworldindata.org/grapher/ghg-per-kg-poore?tab=table, 5 September 2021.

Pimentel, D. (2009). Energy inputs in food crop production in developing and developed nations. *Energies 2*(1), 1–24.

Pimentel, D. and Pimentel, M. (2003). Sustainability of meat-based and plant-based diets and the environment. *The American Journal of Clinical Nutrition*, *78*(3), 660S–663S.

Poore, J. and Nemecek, T. (2018). Reducing food's environmental impacts through producers and consumers. *Science 360*(6392), 987–992.

Porter, S., Reay, D., Higgins, P. and Bomberg, E. (2016). A half-century of production-phase greenhouse gas emissions from food loss and waste in the global food supply chain. *Science of The Total Environment 571*, 721–729.

Raghu, K. (2015). Comparative Lifecycle Assessment on Organic and Conventional Carrots—Case: Carrots from South-Savo and Imported Carrots from Italy. Master in Energy in Technology Thesis. Lappeenranta University of Technology.

The Registrar General & Census Commissioner (2015). Population by Religion Community. Retrieved from https://censusindia.gov.in/2011census/SCST-Series/ST14.html, 5 September 2021.

Schroeder, T.C., Barkley, A.P. and Schroeder, K.C. (1996). Income growth and international meat consumption. *Journal of International Food & Agribusiness Marketing* 7(3), 15–30.

Statista. (2021). Global Business Cities Report Featuring Company DB insights. Retrieved from www.statista.com/news/global-business-cities-reports-2022/en, 2 May 2022.

33R Kenya Project (n.d.). Dairy. Retrieved from www.3r-kenya.org/dairy/, 23 May 2022.

USDA. (2021). FoodData Central. Retrieved from https://fdc.nal.usda.gov/, 12 November 2021.

Whitnall, T., and Pitts, N. (2019). Global trends in meat consumption. *Agricultural Commodities 9*(1), 96–99.

Wolfram|Alpha. (2021). Wolfram|Alpha Examples: Food and Nutrition. Retrieved from www.wolframalpha.com/examples/society-and-culture/food-and-nutrition/, 12 November 2021.

Xu, X., Zhang, B., Liu, Y., Xue, Y. and Di, B. (2013). Carbon footprints of rice production in five typical rice districts in China. *Acta Ecologica Sinica 33*, 227–232.

4

ENERGY

4.1 Introduction

At its simplest, energy is the ability to do work. Humans, being animals, obtain energy from food, while plants get it from sunlight. If you eat plants, or you eat animals that have eaten plants, then indirectly the energy that allows you to work comes from the sun. With the exception of radioactive elements, which can be harnessed for energy in the form of heat in a nuclear reactor, and geothermal energy, which uses the heat in the earth's core, all other energy sources consumed in cities come from the sun. This might be in the form of burning fossil fuels (stored sunlight), the potential energy used in hydropower stations that comes from the sun-driven hydrological cycle, wind energy from the sun and earth's rotation that produces the winds or direct solar energy in the form of heat or converted to electricity. Without the sun, there would be no work done on planet earth.

Because cities are the economic centres of national economies, it should be no surprise that the use of energy in cities is often higher than the national average. Chowdhury et al. (2020) note that although 55% of the global population live in urban areas, these same urban areas consume 75% of global primary energy. Within this simple ratio, things are more complicated when it comes to energy use. In their investigation of large city data sets, Bettencourt and West (2010) found it was size rather than geography or culture that mostly determined the characteristics of a city. So as cities grow, so will their consumption of energy. The same researchers found that three characteristics varied as the population of a city increased.

> One, the space required per capita shrinks, thanks to denser settlement and a more intense use of infrastructure. Two, the pace of all socioeconomic activity accelerates, leading to higher productivity. And three, economic and social

DOI: 10.4324/9781003045847-5

activities diversify and become more interdependent, resulting in new forms of economic specialization and cultural expression.

So, although densification of cities might reduce energy use in terms of transport and residential energy (Newman and Kenworthy, 1989; Norman et al., 2006; O'Brien et al., 2010), it will lead to greater energy use through increased productivity and additional activities that might not be available to rural populations. From this, it might be expected that there is a relationship between the wealth of cities and the energy used within them. This will be explored in this chapter.

4.2 Measuring energy consumption

The first issue in dealing with energy consumption is to distinguish between primary and delivered energy. Delivered energy is measured at the point of energy use. It is what is recorded on your electricity meter or stacked in your wood shed. Primary energy includes all the energy to extract or process fuel and deliver it. In the case of electricity, as well as the fuel, it includes the energy used to build the power station and the energy to make the transmission cables and pylons as well as losses in those cables and in the wires used to transmit the power from the plant to the plug.

In this chapter, the intention is to deal with primary energy. There is an additional complication in that not all energy consumed in cities is necessarily national energy. Nearly all the cities are locked in a permanent electricity exchange, the grid getting and providing energy to compensate for the difference in demand for different places and hours. It is possible that New York, in some parts of the day, is using energy from California, or even from Canada or Mexico. This makes it hard to say what the carbon emissions for New York's electricity are. Electricity in California (2016) had an emissions factor for generation and delivery of 0.22 $kgCO_2/kWh$, while that for Canada (2017) was 0.14 $kgCO_2/kWh$, and for Mexico (2017) was 0.46 $kgCO_2/kWh$ (Carbon Footprint, 2019). There is another issue of dates, since how electricity is generated can vary from year to year. For example, over the five years between 2016 and 2020 in New Zealand, the share of renewables in total primary energy was 39% (2016), 38% (2017), 40% (2018), 39% (2019) and 40% (2020). In 2020, of the total supply of renewables, geothermal accounted for 57%, hydro 25% and solid biofuels 14%, the latter including firewood burned for heating (Ministry of Business, Innovation and Employment, 2020). For the same dates, the share of renewables in New Zealand electricity was 85% (2016), 82% (2017), 84% (2018), 82% (2019) and 81% (2020) (Statista, 2021). In 2017 there was lower than normal rainfall in parts of the South Island affecting the supply of hydro power, this coupled with an expanding population in New Zealand and hence a rising demand for energy and electricity underlines the problem of comparing energy statistics, which normally vary from year to year. Averages over several years could give a better indication, although population increases over time and with it, energy demand.

4.3 Energy and climate change

The importance of energy is intrinsically linked to the problem of fossil fuels and climate change. In 2017 the European Energy Agency (EEA) stated:

> Globally, the use of energy represents by far the largest source of greenhouse gas emissions from human activities. About two thirds of global greenhouse gas emissions are linked to burning fossil fuels for energy to be used for heating, electricity, transport and industry. In Europe, too, the energy processes are the largest emitter of greenhouse gases, being responsible for 78% of total EU emissions in 2015.

Burning any fossil fuel releases carbon dioxide (CO_2), which continues to be the largest driver of human-induced climate change (IPCC, 2014:135). However, the fossil fuel industry also releases methane, a more potent greenhouse gas, during mining and extraction processes. This is estimated to be around 15–20% of all methane emissions, the remaining much bigger contribution coming from agriculture. However, a recent study suggests this is an underestimate and that "total fossil fuel [methane] emissions ... are not increasing over time, but are 60 to 110 per cent greater than current estimates" (Schwietzke et al., 2016). In some ways, this is good news, since if humanity stops using fossil fuels, then not only is CO_2 avoided but also at least a fifth of global methane production.

Johnsson et al. (2019) state that despite investments in renewable energy systems, fossil fuels still provide around 80% of global energy, commenting that simultaneously "large reserves of fossil fuels, especially coal, remain unexploited." The same authors also note that nations with large fossil fuel reserves, such as the Middle East, China, India and Venezuela, have seen an increase in primary energy from fossil fuels, while this has not happened in nations with small reserves such as Brazil, Japan and Canada.

To see whether this applies to the countries where the 19 cities are situated, Table 4.1 presents the national coal and oil reserves of the nations of the 19 cities as well as the potential for these fuels to contribute to global warming if they are consumed. The chief global warming potential from coal comes from the CO_2 released when it is burned. The US Energy Information Administration (EIA) (2021) states that coal mining releases methane, which is a more potent greenhouse gas by a factor of 25. "In 2019, methane emissions from coal mining and abandoned coal mines accounted for about 8% of total U.S. methane emissions and about 1% of total U.S. greenhouse gas emissions (based on global warming potential)." Methane has been ignored in the simple calculation in Table 4.1, which focuses solely on CO_2 and how much of this gas might be produced if all fossil fuel reserves were to be consumed.

According to the New Zealand National Institute of Water and Atmosphere (NIWA, n.d.) for every 1 million tonnes of coal burned in a power station, around 2.5 million tonnes of CO_2 are emitted into the atmosphere (unless deliberately

TABLE 4.1 National global warming potential in 2016 from known reserves of coal and oil

Country	*Coal million tons 2016*[1]	*CO_2 million tonnes from burning coal*	*Oil reserves Million barrels 2016*[2]	*CO_2 million tonnes from burning oil*	*Natural gas billion cubic metres*[3] *2015/16*[4]	*CO_2 million tonnes from burning gas*
United States	245,197	612,993	35,230	15,149	8,630	17,260
Russia	167,771	419,428	80,000	34,400	50,485	100,970
Australia	159,634	399,085	1,193	513	3,791	7,582
China	149,818	374,545	25,132	10,807	2,644	5,288
India	107,727	269,318	4,729	2,033	1,489	2,978
Germany	39,802	99,505	133	57	42	84
Indonesia	24,910	62,275	3,693	1,588	2,775	5,550
New Zealand	8,350	20,875	64	28	61	122
Brazil	7,271	18,178	16,184	6,959	429	858
North Korea	661	1,653	0	0	0	0
Argentina	551	1,378	2,380	1,023	325	650
Netherlands	548	1,370	141	61	732	1,464
Malaysia	198	495	3,600	1,548	2,740	5,480
Egypt	18	45	4,400	1,892	2,086	4,172
Kenya	0	0	0*	0	0	0
Denmark	0	0	551	237	79	158
UAE	0	0	97,800	42,054	6,091	12,182
Cuba	0	0	124	53	11	22
Singapore	0	0	0	0	0	0

1 Worldometer, (n.d.a).
2 Worldometer (n.d.b).
3 eni spa, 2016:38–40).
4 As at 31 December 2015.
* Ngugi (2021) states Kenya has since discovered oil reserves of 2.85 billion barrels of which more than 80% is not recoverable with current extraction technology.

captured). Given the small difference in tons and tonnes, a factor of 2.5 has been used in Table 4.1 for the conversion of tons of coal to tonnes of CO_2. According to the US Environmental Protection Agency (EPA, n.d.) when a barrel of oil is burned, 0.43 tonnes of CO_2 are produced. Natural gas is measured in billion cubic metres, which is equivalent to a cubic kilometre. There are various factors for converting natural gas to CO_2 emissions and no real agreement as to what figure to use since it depends on the particular source of gas and how the gas is burned (Krtkova et al., 2014). According to BP (2021) 1 billion cubic metres of natural gas is equivalent to 36 PJ (36,000 TJ). Krtkova et al. (2014) state that burning one TJ of natural gas emits 56.1 tonnes of CO_2. This means burning each billion cubic metres of gas would emit 2 million tonnes of CO_2. As a check, this is very similar to the Norwegian figure for natural gas of 1.9 kgCO_2/m^3 (1.9 million tonnes/billion cubic metres) (Innovation Norway, n.d.). For simplicity, the factor of 2 million tonnes CO_2 per billion cubic metres of natural gas has been applied to the final column in Table 4.1. Countries have been ranked in Table 4.1 by coal reserves as globally these are much greater than oil and natural gas reserves.

Table 4.2 indicates the potential CO_2 emissions per head of the 2016 population assuming the coal and oil are consumed rather than being left in the ground (Climate Council of Australia, 2015). This takes no account of where the resources are consumed, given the trade in gas, oil and coal, but does indicate which countries

TABLE 4.2 Potential tonnes CO_2 per head from consuming 2016 fossil fuel reserves

Country	*Total million tonnes CO_2 from all reserves*	*Population 2016 millions*[1]	*Potential CO_2 tonnes per head 2016*
United States	645,402	323.9	1,993
Russia	554,798	144.3	3,845
Australia	407,180	24.1	16,895
China	390,640	1378.0	283
India	274,329	1328.9	206
Germany	99,646	82.6	1,205
Indonesia	69,413	259.4	268
New Zealand	21,025	4.7	4,473
Brazil	25,962	206.1	126
North Korea	1,653	25.1	66
Argentina	3,051	43.6	70
Netherlands	2,895	17.0	170
Malaysia	7,523	30.8	244
Egypt	6,109	93.5	65
Kenya	0	45.4	0
Denmark	395	5.7	69
UAE	54,236	9.3	5,831
Cuba	75	11.2	7
Singapore	0	5.6	0

1 Population Reference Bureau (n.d.).

have the potential to cause a significant contribution to global warming if they do not leave their fossil fuels in the ground.

Table 4.2 highlights Australia as a country with significant coal reserves and a low population, leading to a high potential for CO_2 per person. Also significant in this respect, although to a lesser degree, are Russia, New Zealand and the UAE. At the other extreme are countries like Cuba and Denmark with relatively small populations and small reserves of oil and natural gas but no coal.

Table 4.3 sets out the renewable energy capacity per head for 2016 from all types of renewables (hydro, wind, solar and biomass) (IRENA, 2021a:5–38). To show the changing state of the situation, this is compared to the renewables capacity in 2020 in the final column. Capacity is used here to give an idea of the amount of investment taking place in renewables, as capacity is the amount of electricity generated by a power plant if it is running flat out. The energy generated by renewables will vary from day to day and year to year depending on the weather. A 1 MW wind turbine will not generate 1 MW 24 hours a day. This applies to non-renewables as well: a 1 GW nuclear power station will not produce 1 GW of power 24 hours a day; it will depend on the "capacity factor". Capacity factors reflect actual generation as a percentage of continuous operation at full capacity; they vary from around 73% for nuclear and 55% for coal, to 39% for hydro and 12% for solar (global averages 2008–2012 calculated from US EIA data (EIA, 2015)). Looking at capacity factors, hydro can give a more reliable renewable electricity supply than wind or solar, although drought can lead to energy shortages. Table 4.3 shows that renewables capacity is rising in nearly all the countries where the 19 cities are located.

When renewable capacity per head (*y*-axis) is plotted against potential CO_2 per head (*x*-axis) for 2016 (Table 4.2), there is no obvious relationship (Figure 4.1). Having greater fossil fuel reserves does not appear to determine whether or not a country will invest in renewables.

In 2016 Australia had the fourth highest renewables capacity at 1.4 kW/head, noting countries in Table 4.3 are still ranked by their level of coal reserves as shown in Table 4.1. Three of the top four countries for renewables capacity (Australia,

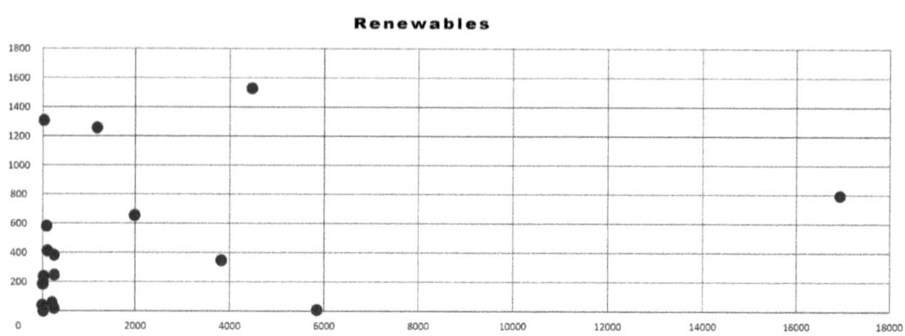

FIGURE 4.1 Fossil fuel reserves and renewables capacity

TABLE 4.3 Renewables capacity per head in 2016 and 2020

Country	*Renewable capacity MW 2016*[1]	*Population 2016 millions*[2]	*Renewable capacity /head (2016) W*	*Renewable capacity MW 2020*[1]	*Population 2020 millions*[3]	*Renewable capacity /head (2020) W*
USA	215, 087	323.9	664	292,065	329.5	886
Russia	51,337	144.3	356	54,274	144.1	377
Australia	19,324	24.1	802	35,690	25.7	1,389
China	541,006	1378.0	393	894,879	1,410.9	634
India	90,313	1328.9	68	134,197	1,380.0	97
Germany	104,436	82.6	1,264	131,739	83.2	1,583
Indonesia	9,064	259.4	35	10,554	273.5	39
New Zealand	7,183	4.7	1,528	7,418	5.1	1,455
Brazil	121,395	206.1	589	150,047	212.6	706
North Korea	4,733	25.1	189	4,823	25.8	187
Argentina	10,670	43.6	245	14,072	44.5	316
Netherlands	7,200	17.0	422	17,678	17.4	1,016
Malaysia	7,958	30.8	258	8,699	32.4	268
Egypt	3,676	93.5	39	5,980	102.3	58
Kenya	1,613	45.4	36	2,191	53.8	41
Denmark	7,410	5.7	1,300	9,677	5.8	1,668
UAE	142	9.3	15	2,540	9.9	257
Cuba	610	11.2	54	1,198	11.3	106
Singapore	307	5.6	55	549	5.7	96

1 IRENA (2021:2–4).
2 Population Reference Bureau (n.d.).
3 World Bank Group (2020).

Germany and New Zealand) fall within the top half of the table for fossil fuel reserves, the exception being Denmark. By 2020 five countries have an installed renewables capacity of more than 1 kW/head, and of these the same three (Australia, Germany and New Zealand) are in the top half of the table for fossil fuel reserves with the Netherlands joining Denmark in the lower half of the table. It appears that it is wealthier countries that are much more able to invest in renewables whatever their fossil fuel reserves. To explore this, Table 4.4 sets out renewable capacity in 2020 with GDP per head for the same year. In Table 4.4 countries are now ranked by their renewable capacity per head.

TABLE 4.4 National GDP and renewable capacity per head for 2020

Country	*Renewable capacity/head 2020 W*	*GDP per head 2020 (US$)*[1]
Denmark	1,668	61,063.3
Germany	1,583	46,208.4
New Zealand	1,455	41,441.5
Australia	1,389	51,692.8
Netherlands	1,016	52,397.1
United States	886	63,413.5
Brazil	706	6,796.8
China	634	10,434.8
Russia	377	10,126.7
Argentina	316	8,579.0
Malaysia	268	10,412.3
UAE	257	36,284.6
North Korea	187	N/A
Cuba	106	9,477.9
India	97	1,927.7
Singapore	96	59,797.8
Egypt	58	3,569.2
Kenya	41	1,878.6
Indonesia	39	3,869.6

1 World Bank Group (2022a).

The big exception to the idea that wealthy nations are more likely to invest in renewables is Singapore with the third highest GDP but ranked 16th for renewables capacity; otherwise it seems that it is not the availability of fossil fuels that is preventing investment in renewables but rather the money or political will to do this. China has only 70% as much "fossil carbon" as the Russian Federation (Table 4.2) but has 168% of the Russian Federation's installed renewables capacity. O'Meara (2020) notes of China that "The country has made massive investments in renewables, depositing 0.9% of its gross domestic product (GDP) into the sector in 2015—the third-highest amount worldwide after Chile and South Africa, which each invested 1.4% of GDP." (None of the cities that form the subject of this book are in Chile or South Africa.)

There is another issue to be explored—that of the renewable capacity per head compared to how much energy each person uses. Table 4.5 gives the MWh per person generated from renewables (hydro, wind, solar, geothermal, wave, tidal and bioenergy but not traditional biofuels such as wood) in 2020, and total primary energy per capita in 2019, the date for which data is available. Countries are still ranked by 2020 renewable capacity and the table reveals the huge gap between delivery from renewables and consumption for some countries, no matter whether the country is rich or poor.

TABLE 4.5 Comparison of energy from renewables and total energy per head for 2019

Country	*Renewables per capita MWh*[1]	*Total primary energy per capita kWh*[2]	*% energy from renew-ables*	*Ranking for % renewable energy*	*GDP ranking (2020)**
Denmark	10.1	33,535	30.1	3	2
Germany	7.6	43,703	17.4	4	6
New Zealand	18.8	53,225	35.3	2	7
Australia	6.1	70,644	8.6	8	5
Netherlands	3.7	57,047	6.5	10	4
United States	7.0	79,897	8.8	7	1
Brazil	7.3	16,325	44.7	1	13
China	3.5	27,425	12.8	6	9
Russia	3.3	56,756	5.8	13	11
Argentina	2.9	21,480	13.5	5	12
Malaysia	2.2	37,054	5.9	12	10
UAE	1.1	137,318	0.8	15	8
North Korea		4,521 (2016)	N/A		
Cuba		11,127 (2016)	N/A		
India	0.5	6,924	7.2	11	16
Singapore	0.4	169,886	0.2	16	3
Egypt	0.5	10,753	4.6	14	15
Kenya		1,906 (2016)	N/A		
Indonesia	0.6	9,147	6.6	9	14

1 Ritchie and Roser (n.d.a).
2 (Ritchie and Roser (n.d.b).
* Omitting Cuba, Kenya and North Korea, rankings taken from Table 4.4.

Both Singapore and the UAE have low investment in renewables and high GDP per capita. At the other extreme is Brazil, ranked 13th for GDP and first for energy from renewables. In 2020, 66% of Brazil's electricity came from hydro (EIA, 2021), mostly located in the Amazon River basin. Argentina is ranked 5th for renewable energy generation and 12th for GDP, but both Argentina and Brazil have a much lower per capita energy use than the six wealthy countries at the top of the table. How much energy is produced from the installed renewables also depends on climate and geography. Not every country can make use of hydro, Denmark being a case in point with only 7 MW installed hydro capacity, effectively 0%, while wind was 64% of Denmark's renewables capacity (IRENA, 2021b). China is ranked sixth for renewable energy capacity and ninth for GDP, thus reflecting the earlier comment that China is deliberately investing GDP in renewables. Indonesia is perhaps a surprise being ranked 9th for capacity and 14th for GDP, with India in a similar situation ranking 11th for renewables capacity and 16th for GDP. Despite appearing to be much poorer countries compared to the wealthy ones in Table 4.4 they are relatively investing more in renewables than the wealthy, whereas it should be the wealthy, developed countries which have had the major share in creating the climate crisis that should be leading the way.

This analysis only concerns the countries where the 19 cities are located and so should be taken as indicative, although suggesting an area that might benefit from a more extensive analysis. However, it does provide some data, which can be compared with the energy consumption of the cities, which is explored in Section 4.5. As noted at the start of this chapter, cities tend to be higher energy consumers than national averages.

4.4 Renewables and greenhouse gas emissions

The tables above looked at the CO_2 produced by fossil fuels but did not consider the CO_2 emissions associated with different types of renewables. The problem with doing this is that the information varies according to who did the analysis. Amponsah et al. (2014) undertook a literature review of peer-reviewed life-cycle energy studies of renewable systems for generating electricity. Concentrating on carbon dioxide and methane, they found a wide range of values for all technologies. As an example, from 14 studies, the range for onshore wind was 1.7–81 gCO_{2eqv}/kWh_e with a mean value of 16.0 gCO_{2eqv}/kWh_e. Most emissions came from extracting the raw materials and manufacturing, shipping and installing the machines. Hydro with 11 studies also displayed a wide range from 2 gCO_{2eqv}/kWh_e to 60 gCO_{2eqv}/kWh_e, with a mean of 20 gCO_{2eqv}/kWh_e. Table 4.6 shows the worst-case values from the literature reviewed by Amponsah et al. (2014), together with the average values for both renewable systems and fossil fuels when generating electricity. It shows that not all renewable technologies are created equal, but all worst cases, with the exception of dedicated biomass waste treatment, and all mean values, are better than mean values for fossil fuels.

Nuclear energy may be associated with low emissions, but it does come with other problems, including waste disposal and the possibility of catastrophic failure, plus uranium is a finite fuel just like coal, oil or gas. The real message from the discussion above is the need to reduce energy demand drastically, leave fossil fuels in the ground and invest in renewables. Table 4.5 reveals the extent of this issue for some countries, with primary energy use per capita ranging from a low of 1,906 kWh/annum (Kenya) to 169,886 kWh/annum (Singapore), nearly 100 times as much. The climate (see Table 1.5) in Nairobi (average temperature 18.8°, annual rainfall 674 mm) is more clement than that of Singapore (average temperature

TABLE 4.6 Life-cycle greenhouse gas emissions and renewable energy technologies (Amponsah et al., 2014)

Technology	*Worst emissions* gCO_{2eqv}/kWh_e	*Mean value emissions* gCO_{2eqv}/kWh_e
Offshore wind	24.0	9.0
Nuclear	24.2	N/A
Tidal	50.0	15.0
Wave	50.0	22.8
Hydro	74.9	20.0
Geothermal	78.0	32.0
Onshore wind	123.7	34.2
Solar thermal	150.0	122.9
Photovoltaics	300.0	77.1
Dedicated biomass	650.0	88.0
Waste treatment	1000.0	399.0
Natural gas	N/A	499.0
Oil	N/A	733.0
Coal	N/A	888.0
Lignite	N/A	1054.0

26.7°, annual rainfall 2366 mm), but it is much more likely to be wealth that is behind such a big difference in energy use.

4.5 Energy use in the cities

The intention in this section was to compare the energy use in cities with national averages, but it soon became very clear that such data was unobtainable, making it necessary to work with national energy consumption in the calculation of the EF. As a first step, Table 4.7 sets out the 2019 national energy mix (where known) by fuel type but excluding biomass. All the figures are from one source for consistency (Ritchie and Roser, 2020). This breakdown is a necessary step in calculating the EF of energy use for the cities. Any differences between national energy use and energy use for the cities for which there is data are discussed below.

In 2019 6.4% of global primary energy came from traditional biomass, defined as "the burning of solid fuels such as wood, crop waste, or charcoal" (Ritchie and Roser, 2020). The same authors also note that traditional biomass can be a very important energy source in some low-income nations, which might apply to Ahmedabad and Nairobi. These issues are examined below where some cities from Table 4.7 are pulled out for further discussion, including those with a relatively high "Not specified" figure and instances where Ritchie and Roser (2020) did not provide any data. For all cities other than those discussed below, the intention was to apply an average value for the impact of all fuels to the "Not specified" category when calculating the energy EF.

TABLE 4.7 Percentage breakdown of national primary energy, excluding traditional biomass, by fuel type 2019 (all data from Ritchie and Roser, 2020) (numbers rounded)

City	*Coal*	*Gas*	*Oil*	*Nuclear*	*Hydro*	*Solar*	*Wind*	*Not specified*
Ahmedabad	54.7	6.3	30.1	1.2	4.2	1.2	1.7	0.7
Amsterdam	7.6	37.8	47.1	1.0	0	1.2	2.9	2.9
Auckland	6.8	18.9	39.0	0	24.9	0.1	2.2	8.2
Beijing	57.6	7.8	19.7	2.2	8.0	1.4	2.6	0.7
Buenos Aires	0.7	49.4	34.2	2.2	9.5	0.2	1.3	2.5
Cairo	2.1	54.6	38.7	0	3.1	0.9	0.6	0
Copenhagen	5.6	15.0	47.8	0	0	1.2	20.7	9.7
Dubai	2.1	56.6	40.5	0	0	0.8	0	0
Hamburg	17.5	24.3	35.6	5.1	1.4	3.2	8.6	4.4
Havana	N/A	N/A	N/A	N/A	N/A	N/A	N/A	N/A
Jakarta	38.2	17.7	38.0	0	1.7	0	0	4.4
Kuala Lumpur	21.1	35.7	36.8	0	5.7	0.2	0	0.6
Moscow	12.2	53.7	22.0	6.3	5.8	0	0	0
Nairobi	N/A	N/A	N/A	N/A	N/A	N/A	N/A	N/A
New York	12.0	32.2	39.1	8.0	2.6	1.0	2.9	2.3
Pyongyang	N/A	N/A	N/A	N/A	N/A	N/A	N/A	N/A
Sao Paulo	5.3	10.4	38.1	1.2	28.7	0	4.0	11.9
Singapore	0.7	12.9	86.1	0	0	0.1	0	0.2
Sydney	27.8	30.2	33.4	0	2.0	2.5	2.7	1.4

4.5.1 *Ahmedabad*

There are no figures for the city, but there are figures for 2018 for the state of Gujarat. Gujarat has the highest energy per capita consumption of all Indian states at 11,630 kWh (IEA, 2021) compared with average Indian primary energy consumption of 6,924 kWh/capita (Table 4.5).The 2018 breakdown of primary energy consumption per capita for Gujarat (IEA, 2021) is coal 52.0%; oil 14.4%; electricity 13.6%; traditional biomass 17.1%; and other 2.9%. Buckley and Shah (2019:12) give the following breakdown for electricity generation in Gujarat for 2018/19: coal 70.1%; gas 11.0%; nuclear 0.9%; hydro 0.9%; renewables 11.8%; and imports 5.3%. Using these percentages to break down the 13.6% of electricity in total energy use gives the following (figures rounded): coal 9.5%; gas 1.5%; nuclear 0.1%; hydro 0.1%; renewables 1.6%; and imports 0.7%. Table 4.8 compares these values with those from Table 4.7. Renewables have been split equally between wind and solar. Although biomass is still an issue for the national values, it seems the state values would give a better picture of energy use in Ahmedabad.

4.5.2 *Auckland*

Among other sources, New Zealand uses geothermal power for generating electricity, something not considered in Table 4.7.The New Zealand Ministry of Business, Innovation and Employment (MBIE) (2020:10–11) gives the 2019 primary energy breakdown as shown in the lower line in Table 4.9, where it is compared with the

TABLE 4.8 Primary energy breakdown for India and Gujarat

City	*Coal*	*Gas*	*Oil*	*Nuclear*	*Hydro*	*Solar*	*Wind*	*Biomass*	*Not specified*
Ahmedabad (national)	54.7	6.3	30.1	1.2	4.2	1.2	1.7	N/A	0.7
(Gujarat)	61.5	1.5	14.4	0.1	0.1	0.8	0.8	17.1	3.6

TABLE 4.9 NZ primary energy breakdown

City	*Coal*	*Gas*	*Oil*	*Hydro*	*Solar*	*Wind*	*Geo-thermal*	*Biomass*	*Waste heat*	*Not specified*
Auckland	6.8	18.9	39.0	24.9	0.1	2.2	N/A	N/A	N/A	8.2
NZ	7.1	20.5	32.8	10.2	0.1	0.9	21.7	6.6	0.1	

values in Table 4.7. Nuclear has been omitted as there is no nuclear power in New Zealand.

The big change in Table 4.9 is the relative balance of geothermal and hydro. Table 4.9 also recognises the contribution of firewood for heating in New Zealand. Auckland Council has rules for the use of wood burners in an urban situation (Auckland Council, 2021), suggesting that wood could be burned in the city.

4.5.3 Copenhagen

In 2019 renewables formed 36% of total Danish energy of which 75% came from biomass and "more than 60% of district heating is now produced from biomass," with combined heat and power plants (CHP) having been switched from fossil fuels to biomass (IEA Bioenergy, 2021). Figures for energy consumed in Copenhagen in 2015 also reveal 49% of energy was classed as district heating, of which 20% was coal, 12% gas, 1% oil, 39% biomass and 28% waste (Global Covenant of Mayors for Climate and Energy, n.d.). The same source gives 9% of electricity as coming from biomass, presumably as part of operating combined heat and power (CHP) plants (Danish Energy Agency, n.d.). From this, it would seem that the "Not specified" category in Table 4.7 could be classed as biomass. This gives 31.6% of fuels coming from renewables as against 36% as stated by IEA Bioenergy (2021), again showing the problem of having agreed and reliable data. Copenhagen is also different from many cities in terms of its transport energy. The breakdown of all trips to, from and in Copenhagen in 2019 are car 30%, bike 28%, walking 21% and public transport 21% (Keim and Cerny, 2021:31). This compares with a 2019 modal split for all journeys in Denmark of car 59.3%, walk 18.7%, bike 14.5% and public transport 7.4% (Christiansen and Baescu, 2020:6). In the absence of further data, this suggests the use of oil for transport in Copenhagen is less than in the whole of Denmark because of the much greater use of bicycles in the city. Consequently, in Table 4.10 renewables have been increased to 36% in the form of additional biomass and oil has been reduced.

TABLE 4.10 Suggested 2019 percentage breakdown for energy use in Copenhagen

	Coal	*Gas*	*Oil*	*Nuclear*	*Hydro*	*Solar*	*Wind*	*Not specified*	*Biomass*
Denmark	5.6	15.1	47.8	0	0.02	1.2	20.7	9.7	N/A
Copenhagen	5.6	14.9	43.5	0	0.02	1.2	20.7	0	14.1

TABLE 4.11 2018 percentage breakdown of primary energy for Cuba

	Gas	*Oil*	*Biomass*	*Solar*	*Wind*	*Hydro*
Cuba	9.0	75.0	15.7	0.2	0.1	0.1

TABLE 4.12 2018 primary energy percentage profile for Kenya

	Coal	*Gas*	*Oil*	*Nuclear*	*Geothermal*	*Biomass*	*Hydro*
Kenya	2.0	0	20.0	0	19.5	57.0	1.5

4.5.4 Havana

The 2018 primary energy profile for Cuba is oil 75%, renewables 16%, and gas 9%, with renewables consisting of bioenergy 98%, solar 1% and wind and hydro 0.5% each (IRENA, 2021b). Table 4.11 amalgamates these figures with numbers rounded.

4.5.5 Nairobi

The 2018 primary energy profile for Kenya was renewables 78%, oil 20% and coal 2%. The renewables consisted of bioenergy 73%, geothermal 25% and hydro 2% (IRENA, 2021b). Table 4.12 amalgamates these figures.

4.5.6 Pyongyang

The 2018 primary energy mix for North Korea consisted of coal 51%, renewables 33% and oil 15%, of which the renewables break down into 54% bioenergy and 46% hydro (IRENA, 2021b). These figures have been combined in Table 4.13.

4.5.7 São Paulo

Brazil is unique as the only country in this sample of 19 where in 2018 renewables formed the highest percentage of primary energy at 47%, followed by oil 36%, gas 10%, coal 6% and nuclear 1% (IRENA, 2021b). Renewables consisted of 71% bioenergy, 25% hydro and 4% wind. These values are combined and compared to those of Table 4.7 for 2019 in Table 4.14.

TABLE 4.13 2018 primary energy percentage profile for North Korea

	Coal	*Oil*	*Biomass*	*Hydro*
North Korea	51.0	15.0	17.8	15.2

TABLE 4.14 2018 primary energy percentage profile for Brazil

	Coal	*Gas*	*Oil*	*Nuclear*	*Hydro*	*Solar*	*Wind*	*Not specified*	*Biomass*
Brazil (Table 4.7)	5.29	10.39	38.14	1.16	28.7	0.4	4.01	11.91	N/A
Brazil (IRENA,2021b)	6.0	10.0	36.0	1.0	11.8	0	1.9	N/A	33.4

TABLE 4.15 Comparison of national and city 2011 primary energy consumption

	Coal	*Gas*	*Oil*	*Nuclear*	*Hydro*	*Biofuels*	*Geo/solar/wind*	*Others*
Brazil 2011	0.9	5.7	45.3	1.6	14.8	31.5	0.3	0
São Paulo 2011	3.0	6.0	53.0	N/A	9.0	28.0	N/A	2.0

The values for the fossil fuels are comparable but those of renewables change with the inclusion of biomass. As a final comparison Table 4.15 sets out the 2011 values for both Brazil and São Paulo (Investe São Paulo, 2013).

As might be expected, the city uses more oil than the national average and fewer renewables in the form of hydro though the use of biofuels is perhaps surprisingly high. "According to the Executive Summary of Renewable Energy, published by the Department of Energy (SE), there are 192 cogeneration plants currently operating in São Paulo, which corresponds to 52% of total enterprises and 53% of the installed capacity in Brazil (2013)" (Investe São Paulo, 2013). Although somewhat historic, these percentages have been used in the comparison in Section 4.5.8.

4.5.8 Energy profile comparison of the selected cities

As a final step, Table 4.16 compares the national average values from Ritchie and Roser (2020), which leaves out biomass, those of IRENA (2021b) for 2018, which include biomass, and the recalculated values for Ahmedabad, Auckland, Copenhagen and São Paulo. The point of doing this was to calculate the EF for each to see the extent of the difference since IRENA publishes an energy profile that includes biomass for all countries and it would be useful to use values from the same source when calculating the energy EFs. Numbers have been rounded so not all rows add to 100%. Anything classed as "other renewables" has been added to the bioenergy column.

TABLE 4.16 Percentage breakdown of primary energy 2018 (numbers rounded) for four cities

City	*Scope*	*Coal*	*Gas*	*Oil*	*Nuc-lear*	*Hyd-ro*	*Bio*	*Solar*	*Wind*	*Geo-thermal*	*Reference*
Ahmedabad	National	56	6	29	1	4	0.5	1	2	0	R&R
Ahmedabad	National	47	6	25	1	1	20	0.4	1	0	IRENA
Ahmedabad	Gujarat	62	2	14	0.1	0.1	21	1	1	0	IEA/Buckley&Shah
Auckland	National	7	19	41	0	27	-	0.4	2	4	R&R
Auckland	National	5	17	30	0	10	6	0	1	31	IRENA
Auckland	National	7	21	33	0	10	7	0.1	1	22	MBIE
Copenhagen	National	10	17	49	0	<1	3	1	19	0	R&R
Copenhagen	National	14	16	36	0	0	26	0.4	8	0	IRENA
Copenhagen	City	6	15	44	0	0	14	1	21	0	Adjusted
São Paulo	National	6	11	40	1	30	9	0.2	4	0	R&R
São Paulo	National	6	10	36	1	12	33	0	2	0	IRENA
São Paulo	City	3	6	53	0	9	28	0	2	0	ISP

TABLE 4.17 Possible 2018 EFs for the four selected cities

City	*Scope*	*Primary energy (TJ)*[1]	*2018 national population*[2]	*EF/capita gha*	*Reference*
Ahmedabad	National	39,230,523	1,352,642,283	2.00	R&R
	National	39,230,523	1,352,642,283	1.78	IRENA
	Gujarat	39,230,523	1,352,642,283	1.87	IEA/Buckley&Shah
Auckland	National	937,728	4,743,131	9.32	R&R
	National	937,728	4,743,131	6.89	IRENA
	National	937,728	4,743,131	6.27	MBIE
Copenhagen	National	717,179	5,757,131	21.33	R&R
	National	717,179	5,757,131	20.15	IRENA
	National	717,179	5,757,131	18.85	Adjusted
São Paulo	National	12,267,118	209,469,320	3.36	R&R
	National	12,267,118	209,469,320	3.15	IRENA
	City (2011)	12,267,118	209,469,320	3.60	ISP

1 IRENA (2021b).
2 PopulationPyramid, (2019).

Using the 2018 values of total primary energy supply from IRENA (2021b), Table 4.17 sets out the range of possible EFs for the four selected cities for 2018 based on the different percentage allocations for energy sources from Table 4.16.

The first thing to note from Table 4.17 is that for Ahmedabad and São Paulo, the national IRENA values, which include biomass, give the lowest EF and for Auckland and Copenhagen the IRENA figures give results that lie nearer the middle. IRENA has therefore been used as the data source for the ecological footprint values for the 19 cities. Table 4.18 gives the breakdown of national primary energy from IRENA, which is used to generate EFs in Table 4.19. As expected, in Table 4.17 in 2011 São Paulo used more fossil fuels and less renewable energy than the national value. Consequently, the greenhouse gas emissions, which are used to calculate the energy EF, will be higher. This means the final EFs in Table 4.19 based on national data will probably be an underestimate for each city. Values for Copenhagen with its high use of renewables are unexpectedly high, but this is in part a function of the conversion to gha as explained in Chapter 2. The values in Table 4.19 are for primary energy, including that bound up in transport, food, the built environment and all goods that are made and exported.

Although there are tables of worldwide per capita electricity use (World Bank Group, 2022b), there is no similar collected record of worldwide domestic primary energy consumption, the energy used in the home for heating, lighting, cooking and other appliance use. This household energy is what might be considered as the personal share of energy use, or the energy that people have a choice over using, just

TABLE 4.18 Percentage breakdown of 2018 national primary energy (numbers rounded and adjusted so rows add to 100%) for 19 cities (IRENA, 2021b)

City	*Coal*	*Gas*	*Oil*	*Nuclear*	*Hydro*	*Bio*	*Solar*	*Wind*	*Geo-thermal*	*% renewables*★
Ahmedabad	47	6	25	1	1	19	neg★★	1	0	21
Amsterdam	14	42	37	1	0	4	1	1	0	6
Auckland	6	17	30	0	10	6	0	1	31	48
Beijing	62	7	20	2	4	3	1	1	0	9
Buenos Aires	2	54	33	2	4	5	neg	neg	0	9
Cairo	0	50	41	0	2	7	neg	neg	0	9
Copenhagen	14	16	36	0	0	26	neg	8	0	34
Dubai	3	15	81	0	0	neg	1	0	0	1
Hamburg	23	24	32	6	1	9	1	4	neg	15
Havana	0	9	75	0	neg	16	neg	neg	0	16
Jakarta	31	15	32	0	1	16	0	0	5	22
Kuala Lumpur	24	40	32	0	3	1	neg	0	0	4
Moscow	16	55	20	7	2	neg	0	0	neg	2
Nairobi	2	0	20	0	1	57	0	0	20	78
New York	15	32	36	10	1	5	neg	1	neg	7
Pyongyang	51	0	15	0	16	18	0	0	0	34
São Paulo	6	10	36	1	12	33	0	2	0	47
Singapore	3	35	60	0	0	2	neg	0	0	2
Sydney	34	25	34	0	1	4	1	1	0	7

★ Includes geothermal but excludes nuclear.

★★ The energy source is there but the share is very small and lost in rounding.

TABLE 4.19 2018 national primary energy EF per capita

City	*Primary energy TJ*	*2018 national population*	*EF/capita gha*
Ahmedabad	39,230,523	1,352,642,283	1.78
Amsterdam	3,031,444	17,059,560	19.61
Auckland	937,728	4,743,131	6.89
Beijing	130,236,823	1,427,647,789	5.18
Buenos Aires	3,280,908	44,361,150	2.32
Cairo	3,734,509	98,423,602	0.46
Copenhagen	717,179	5,757,131	20.15
Dubai	2,658,279	9,630,966	6.09
Hamburg	12,676,575	83,124,413	17.04
Havana	374,870	11,338,146	0.97
Jakarta	10,466,599	267,670,549	2.25
Kula Lumpur	3,830,217	31,528,033	9.67
Moscow	33,244,594	145,734,034	4.18
Nairobi	979,607	51,392,570	0.13
New York	93,338,524	327,096,263	10.49
Pyongyang	259,121	25,549,606	0.46
São Paulo	12,267,118	209,469,320	3.15
Singapore	1,069,998	5,757,503	52.81
Sydney	5,402,609	24,898,153	3.32

as they have a choice over transport energy, rather than the energy bound up the whole economy. In 2019 in the European Union, this household energy formed 26% of final energy consumption (Eurostat, 2022a), although this is delivered rather than primary energy. In 2020 households used 18.5% of gross available energy (presumably primary energy) (Eurostat, 2022b). This does demonstrate that individual efforts to save energy can only go so far because of the energy bound up in a functioning economic society. Figure 4.2, which sets out the relationship between 2015 energy use per person and GDP, also shows that not all economic activity is equal when it comes to energy.

Although the general trend is that as GDP increases so does energy use, the group of Australia, the Netherlands, Germany and Denmark shows that energy use can vary considerably for approximately the same value of GDP per capita. From Table 4.18, the 2018 renewables components were 7% (Australia), 6% (Netherlands), 15% (Germany) and 34% (Denmark). Although only part of the story, moving to a much greater renewables component should help to decouple energy from GDP. Guo et al. (2021) investigated such decoupling and noted the following. "To better decouple economic growth from energy use, we propose policies for more structural reforms, a clean energy system, improved energy efficiency, and efficient energy demand-side management." Moving to a knowledge-based economy could also reduce national energy use as manufacturing energy disappears, but if it moves to another location there will be no effect unless the

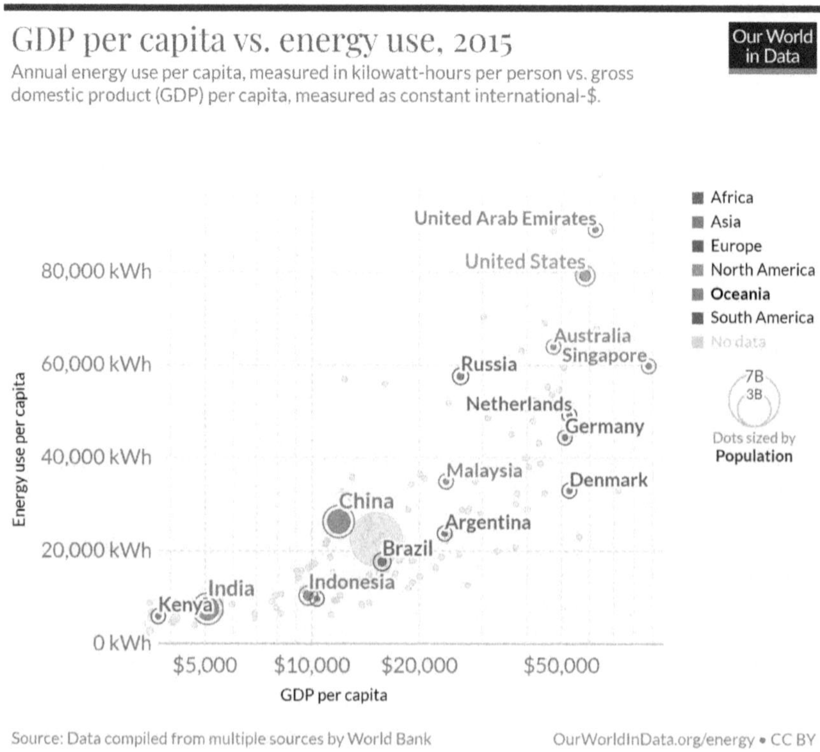

FIGURE 4.2 Energy use and GDP per capita, 2015 (Ritchie and Roser, n.d.c)

demand for manufactured goods also wanes. However, the key remains moving to renewable energy and Table 4.18 indicates how far the world still has to go to achieve this goal.

References

Amponsah, N.Y., Troldborg, M., Kington, B., Aalders, I. and Hough, R.L. (2014). Greenhouse gas emissions from renewable energy sources: A review of lifecycle considerations. *Renewable and Sustainable Energy Reviews 39*, 461–475.

Auckland Council (2021). Rules for domestic heating appliance. Retrieved from www.aucklandcouncil.govt.nz/building-and-consents/building-renovation-projects/heat-your-home/Pages/rules-domestic-heating-appliances.aspx#:~:text=Rules%20and%20standards&text=%E2%80%8BEnclosed%20woodburners%20must%20meet,Other%20domestic%20fires%20permitted.&text=%E2%80%8BAll%20domestic%20solid%20fuel,clause%207%20of%20the%20AQB), 22 February 2022.

Bettencourt, L and West, G. (2010). A unified theory of urban living. *Nature 467*, 912–913.

BP (2021). Approximate conversion factors. Retrieved from www.bp.com/content/dam/bp/business-sites/en/global/corporate/pdfs/energy-economics/statistical-review/bp-stats-review-2021-approximate-conversion-factors.pdf, 9 February 2022.

Buckley, T. and Shah, K. (2019). Gujarat's Electricity Sector Transformation. Retrieved from https://ieefa.org/wp-content/uploads/2019/08/Gujarat-Electricity-Sector-Transformation_Final_August-2019.pdf, 17 February 2022.

Carbon Footprint (2019). Country Specific Electricity Grid Greenhouse Gas Emission Factors. Retrieved from www.carbonfootprint.com/docs/2019_06_emissions_factors_sources_for_2019_electricity.pdf, 9 December 2021.

Chowdhury, P.K.R., Weaver, J.E., Weber, E.M., Lunga, D. LeDoux, St.T.M., Rose, A.N. and Bhaduri, B.L. (2020). Electricity consumption patterns within cities: application of a data-driven settlement characterization method. *International Journal of Digital Earth 13*(1), 119–135.

Christiansen, H. and Baescu, O. (2020). The Danish National Travel Survey—Annual Statistical Report 2019. Retrieved from https://backend.orbit.dtu.dk/ws/files/245076078/TU_Denmark_2019.pdf, 23 February 2022.

Climate Council of Australia (2015). Unburnable carbon: why we need to leave fossil fuels in the ground. Retrieved from www.climatecouncil.org.au/uploads/a904b54ce67740c4b4ee2753134154b0.pdf, 8 February 2022.

Danish Energy Agency (n.d.). Facts about bioenergy in Denmark. Retrieved from https://ens.dk/en/our-responsibilities/bioenergy/facts-about-bioenergy-denmark, 23 February 2022.

EIA (US Energy Information Administration) (2015). Electric generator capacity factors vary widely across the world. Retrieved from www.eia.gov/todayinenergy/detail.php?id=22832, 17 February 2022.

EIA (US Energy Information Administration) (2021). Hydropower made up 66% of Brazil's electricity generation in 2020. Retrieved from www.eia.gov/todayinenergy/detail.php?id=49436, 14 February 2020.

Energy Information Administration (EIA) (2021). Coal explained. Retrieved from www.eia.gov/energyexplained/coal/coal-and-the-environment.php, 4 February 2022.

Eni spa (2016). World Oil and Gas Review 2016. Retrieved from www.eni.com/assets/documents/WOGR-2016.pdf, 8 February 2022.

Environmental Protection Agency (EPA) (n.d.). Greenhouse Gases Equivalencies Calculator—Calculations and References. Retrieved from www.epa.gov/energy/greenhouse-gases-equivalencies-calculator-calculations-and-references, 3 February 2022.

European Energy Agency (2017). Energy and Climate Change. Retrieved from www.eea.europa.eu/signals/signals-2017/articles/energy-and-climate-change, 24 January 2022.

Eurostat (2022a). Energy consumption in households. https://ec.europa.eu/eurostat/statistics-explained/index.php?title=Energy_consumption_in_households, 22 March 2022.

Eurostat (2022b). Energy statistics—an overview. Retrieved from Energy statistics - an overview - Statistics Explained (europa.eu), 30 March 2022.

Global Covenant of Mayors for Climate and Energy (n.d.). Copenhagen municipality. Retrieved from https://dataportalforcities.org/european-union-western-europe/denmark/region-hovedstaden/copenhagen-municipality, 23 February 2022.

Guo, J., Li, C-Z. and Wei, C. (2021). Decoupling economic and energy growth: aspiration or reality? *Environmental Research Letters 16*(4). 44017.

IEA (2021). Final energy consumption per capita in India by state, 2018. Retrieved from www.iea.org/data-and-statistics/charts/final-energy-consumption-per-capita-in-india-by-state-2018, 17 February 2022.

IEA Bioenergy (2021). Implementation of bioenergy in Denmark—2021 update. Retrieved from www.ieabioenergy.com/wp-content/uploads/2021/11/CountryReport2021_Denmark_final.pdf, 23 February 2022.

Innovation Norway (n.d.). Conversion Guidelines—Greenhouse gas emissions. Retrieved from www.eeagrants.gov.pt/media/2776/conversion-guidelines.pdf, 9 February 2022.

Investe São Paulo (2013). Renewable energy. Retrieved from www.en.investe.sp.gov.br/uploads/midias/documentos/renewable_energy_saopaulo.pdf, 25 February 2022.

IPCC (Intergovernmental Panel on Climate Change) (2014). *Climate Change 2014 Mitigation of Climate Change*. Cambridge: Cambridge University Press.

IRENA (International Renewable Energy Agency) (2021a). *Renewable Capacity Statistics 2021*. Abu Dhabi: The International Renewable Energy Agency.

IRENA (2021b). Statistical Profiles. Retrieved from Statistical Profiles (irena.org), 14 February 2022.

IRENA (2021c). Energy Profile: Kenya. Retrieved from www.irena.org/IRENADocuments/Statistical_Profiles/Africa/Kenya_Africa_RE_SP.pdf, 23 February 2022.

Johnsson, F., Kjärstad, J. and Rootzén, J. (2019). The threat to climate change mitigation posed by the abundance of fossil fuels. *Climate Policy 19*(2), 258–274.

Keim, M. and Cerny, P. (2021). *The European Mobility Atlas*. Brussels: Heinrich-Böll-Stiflung.

Krey V., O. Masera, G. Blanford, T. Bruckner, R. Cooke, K. Fisher-Vanden, H. Haberl, E. Hertwich, E. Kriegler, D. Mueller, S. Paltsev, L. Price, S. Schlömer, D. Ürge-Vorsatz, D. van Vuuren, and T. Zwickel. (2014:1297). Annex II: Metrics & Methodology. In *Climate Change 2014: Mitigation of Climate Change. Contribution of Working Group III to the Fifth Assessment Report of the Intergovernmental Panel on Climate Change,* (eds.), Edenhofer, O., R. Pichs-Madruga, Y. Sokona, E. Farahani, S. Kadner, K. Seyboth, A. Adler, I. Baum, S. Brunner, P. Eickemeier, B. Kriemann, J. Savolainen, S. Schlömer, C. von Stechow, T. Zwickel and J.C. Minx. Cambridge: Cambridge University Press.

Krtkova, E, Pavel, Fott and Neuzil, V. (2014). Carbon dioxide emissions from natural gas combustion—country-specific emission factors for the Czech Republic. *Greenhouse Gas Management and Measurement 4*(1), 28–40.

Ministry of Business, Innovation and Employment (2020). *Energy in New Zealand 20.* Wellington: Ministry of Business, Innovation and Employment.

Ministry of Business, Innovation and Employment (2020). *Energy in New Zealand.* Retrieved from www.mbie.govt.nz/dmsdocument/11679-energy-in-new-zealand-2020, 22 March 2022.

Ministry of Business, Innovation and Employment (2020). Renewables statistics. Retrieved from www.mbie.govt.nz/building-and-energy/energy-and-natural-resources/energy-statistics-and-modelling/energy-statistics/renewables-statistics/, 24 January 2021.

National Institute of Water and Atmosphere (NIWA) (n.d.). How can carbon emissions be weighed? Retrieved from https://niwa.co.nz/atmosphere/faq/how-can-carbon-emissions-be-weighed, 3 February 2022.

Newman, P.W.G. and Kenworthy, J.R. (1989). Gasoline consumption and cities. *Journal American Planning Association 55*, 24–37.

Ngugi, B. (2021). 80pc of Kenya's crude oil cannot be tapped. Retrieved from www.businessdailyafrica.com/bd/economy/80pc-of-kenya-s-crude-oil-cannot-be-tapped-3553040, 26 January 2022.

Norman, J., MacLean, H. and Kennedy, C. (2006). Comparing high and low residential density: Life-cycle analysis of energy use and greenhouse gas emissions. *Journal of Urban Planning and Development 132*, 10–21.

O'Brien, W.T., Kennedy, C.A., Athienitis, A.K. and Kesik, T.J. (2010). The relationship between net energy use and the urban density of solar buildings. *Environment and Planning B: Planning and Design 37*, 1002–1021.

O'Meara, S. (2020). China's plan to cut coal and boost green growth. Retrieved from www.nature.com/articles/d41586-020-02464-5, 14 February 2022.

PopulationPyramid (2019). List of countries ordered by their population. Retrieved from Countries Ordered by Population in 2018 (populationpyramid.net), 27 February 2022.

Population Reference Bureau (n.d.). Retrieved from www.prb.org/wp-content/uploads/2016/08/prb-, 25 January 2022.

Ritchie, H. and Roser, M. (2020). Energy. Retrieved from https://ourworldindata.org/energy, 18 February 2022.

Ritchie, H. and Roser, M. (n.d.a). Our World in Data. Retrieved from https://ourworldindata.org/grapher/per-capita-renewables, 14 February 2022.

Ritchie, H. and Roser. M (n.d.b). Our World in Data. Retrieved from https://ourworldindata.org/grapher/per-capita-energy-use, 14 February 2022.

Ritchie, H and Roser, M. (n.d.c). Our World in Data. Retrieved from https://ourworldindata.org/grapher/energy-use-per-capita-vs-gdp-per-capita?zoomToSelection=true&time=2015, 22 March 2022.

Schwietzke, S., Sherwood, O.A., Bruhwiler, L.M.P., Miller, J.B., Etiope, G., Dlugokencky, E.J., Englund Michel, S., Arling, V.A., Vaughn, B.H., White, J.W.C. and Tans, P.P. (2016). Upward revision of global fossil fuel methane emissions based on isotope database. *Nature 538*(7623), 88–91.

Statista (2021). Share of electricity generated from renewable energy in New Zealand from 2011 to 2020. Retrieved from www.statista.com/statistics/987812/new-zealand-renewable-energy-generation-share/, 24 January 2022.

World Bank Group (2022a). GDP per capita, (current US$). Retrieved from https://data.worldbank.org/indicator/NY.GDP.PCAP.CD, 8 February 2022.

World Bank Group (2022b). Electric Power Consumption (kWh per capita). Retrieved from https://data.worldbank.org/indicator/EG.USE.ELEC.KH.PC, 22 March 2022.

Worldometer (n.d.a). Coal reserves by country. Retrieved from www.worldometers.info/coal/coal-reserves-by-country/, 25 January 2022.

Worldometer (n.d.b). Oil reserves by country. Retrieved from www.worldometers.info/oil/oil-reserves-by-country/, 26 January 2022.

5

TRANSPORT

5.1 The cities and transport

Before looking at the individual cities, it makes sense to look at them as a group. In Table 5.1 the cities are grouped by the gross national income (GNI) of the nations in which they are situated (Conceição, 2020:343–345). The nations are in three groups related to income, with the wealthiest at the top. Table 5.1 also shows the number of motor vehicles per person in each nation (Our World in Data, n.d.a). The definition of motor vehicles here comprises all powered road vehicles except motorcycles and other two-wheelers. The averages show that in general, the higher the income, the more motor vehicles.

Using the same categories, we can look at the carbon dioxide equivalent emissions related to transport in the various nations, split between land transport, which tends to be more local, and aviation and shipping, which in many countries will be more likely to have an international aspect (Table 5.2).

As in Table 5.1, Table 5.2 shows that in general richer countries have higher transport emissions. They have more motor vehicles, more motor vehicle emissions and more aviation and shipping emissions. Singapore, the wealthiest nation and the only one which is also a city, has lower motor vehicle emissions than other wealthy countries but very large aviation and shipping emissions, which is not surprising given its high-density nature and its role as a transport and shipping hub. The United Arab Emirates and the Netherlands are in similar, but less extreme, situations. Because of its extreme emissions situation, Singapore has been removed from the averaging in Table 5.2. The land transport emissions of the new world countries (the United States, Australia and New Zealand) are noticeably higher than those of the old-world European countries, as are their motor vehicle figures. Malaysia, one of the middle-income countries, has overall transport emissions the same as wealthier Germany, but in general, wealth equates to higher transport emissions and

DOI: 10.4324/9781003045847-6

TABLE 5.1 2014 national motor vehicles/capita and GNI

City	*Nation*	*GNI*	*Motor Vehicles/Capita by Nation (2014)*
Higher income			**Average 534**
Singapore	Singapore	88,155	149
Dubai	UAE	67,462	313
New York	United States	63,826	797
Copenhagen	Denmark	58,662	480
Amsterdam	Netherlands	57,707	528
Hamburg	Germany	55,314	572
Sydney	Australia	48,085	717
Auckland	New Zealand	40,799	712
Middle income			**Average 260**
Kuala Lumpur	Malaysia	27,534	361
Moscow	Russian Federation	26,157	293
Buenos Aires	Argentina	21,190	314
Beijing	China	16,057	83
São Paulo	Brazil	14,263	249
Lower income			**Average 33**
Cairo	Egypt	11,466	45
Jakarta	Indonesia	11,459	60
Havana	Cuba	8,621	38
Ahmadabad	India	6,681	18
Nairobi	Kenya	4,244	24
Pyongyang	North Korea	N/A	11

to transport emissions that form a higher percentage of total emissions. As Table 5.2 shows, the wealthier nations tend to have higher aviation and shipping emissions than others, which goes some way to explaining why transport emissions form a greater percentage of their total emissions.

5.2 Travel modes

The emissions related to transport depend on the interaction between two aspects of travel: the type of vehicle and the distance travelled. The type of vehicle, what transport planners call the mode, makes a huge difference to emissions, as shown in Figure 5.1, which gives the CO_2 equivalent emissions for one person travelling one kilometre in the UK in 2018.

The carbon emissions of transport modes depend on the occupancy. This is particularly true for public transport vehicles. While a rush-hour bus, with all seats full and passengers standing, will be highly efficient in emissions per passenger, the same will not be true at 2:30 in the afternoon when there are only three passengers and the driver, but the bus still has to run to provide a service. Even at busy times, the bus will be less full at the start of its route if it starts far away from the city centre. This is discussed below in the consideration of the city of Havana.

TABLE 5.2 Transport emissions compared with total emissions in the various nations

City	*Nation: Ranked by Gross National Income (Table 1.8)*	*Motor vehicles per capita by nation*[1]	*Land Transport CO_{2eqv} per person: tonnes*[2]	*Aviation and Shipping per person: tonnes*[3]	*Total transport emissions per person: tonnes*	*Total emissions per person: tonnes*[4]	*Total transport emissions as % of total*
Higher income		**Average 534**	**Average 2.9 tonnes**	**Average**	**Average (without Singapore) 5.2 tonnes**	**Average (without Singapore) 16.7 tonnes**	**Average (without Singapore) 31%**
Singapore	Singapore	149	1.1	30.9	32.0	11.2 **42.1*	287%
Dubai	UAE	313	3.7	8.4	12.1	27.7	44%
New York	United States	797	5.3	0.4	5.7	18.3	31%
Copenhagen	Denmark	480	2.1	0.9	3.0	8.4	35%
Amsterdam	Netherlands	528	1.8	3.0	4.8	11.0	43%
Hamburg	Germany	572	1.9	0.4	2.3	10.3	22%
Sydney	Australia	717	4.0	0.6	4.6	24.1	19%
Auckland	New Zealand	712	3.1	0.8	3.9	17.2	23%
Middle income		**Average 260**	**Average 1.3**		**Average 1.4**	**Average 9.7**	**Average 13%**
Kuala Lumpur	Malaysia	361	2.0	0.3	2.3	9.5	24%
Moscow	Russian Federation	293	1.7	0.3	2.0	16.7	12%
Buenos Aires	Argentina	314	1.1	0.1	1.2	8.7	2%
Beijing	China	83	0.6	0.04	0.6	8.4	8%
São Paulo	Brazil	249	1.0	0.08	1.1	5.1	21%

Lower income		**Average 33**	**Average 0.3**		**Average 0.3**	**Average 2.7**	**Average 12%**
Cairo	Egypt	45	0.6	0.02	0.6	3.3	19%
Jakarta	Indonesia	60	0.5	0.01	0.5	3.3	16%
Havana	Cuba	38	0.1	0.2	0.3	3.6	9%
Ahmadabad	India	18	0.2	0.02	0.2	2.3	9%
Nairobi	Kenya	24	0.2	0.04	0.2	1.6	14%
Pyongyang	North Korea	11	0.1	n/a	0.1	2.0	3%

* Total emissions for Singapore plus Singapore Aviation and Shipping.
1 Our World in Data (n.d.a).
2,3 Calculated from Ritchie and Roser (2020).
4 Our World in Data (n.d.b).

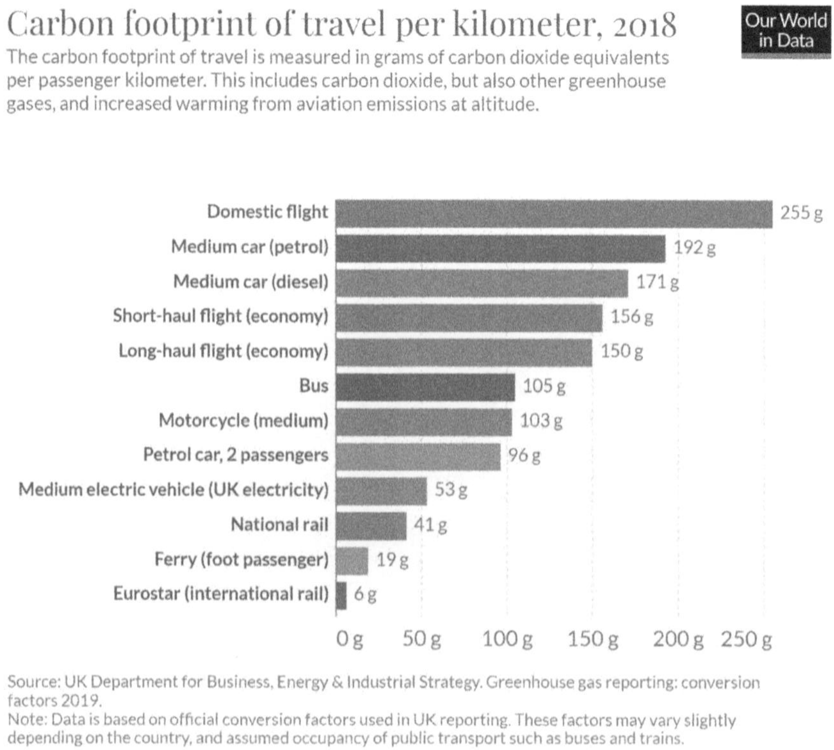

FIGURE 5.1 Carbon emissions of transport modes in grams of CO_2 equivalent per passenger-kilometre (Ritchie, 2020)

The details of the vehicle also matter, not just the mode. Figure 5.1 shows emissions for a medium car at 192 gCO_{2eq} per km. The same source cites a relatively small Mini as having an emission of 111 gCO_{2eq} per km while a large four-by-four sport utility vehicle (SUV) will be around 200 gCO_{2eq} per km (Ritchie, 2020). Not shown in the chart in Figure 5.1 is cycling as a mode of transport. This is quoted as having a carbon footprint between 16 and 50 grams CO_{2eq} per km depending partly on what you eat to power your bike ride (Ritchie, 2020).

Some of the data in Figure 5.1 will apply to all countries. Any vehicle (car, bus, train, aircraft, ferry, motorbike) using petrol or diesel will have the same emissions anywhere because the fuel is the same. This is not true of vehicles powered by electricity because the mix of fuels used to generate electricity differs from one country to another. Emissions even differ within a single country. The US Department of Energy (n.d.) shows emissions of electric, plug-in hybrid, hybrid and all-petrol cars state-by-state in the United States as well as nationally. In states like Vermont, which generate most of their electricity from renewable sources such as hydroelectricity, the electric car can have zero emissions, while in West Virginia, where 88% of electricity comes from burning coal, the hybrid petrol car has the lowest emissions. The same issue applies to electrically powered transport in different countries, depending on how electricity is generated.

In this chapter, the emissions for oil-fuelled vehicles and bicycles are based on the data from Ritchie (2020) and the emissions for electrically powered transport are calculated for each case according to how electricity is generated in the nation, or the state, where the city is located.

5.3 Travel distances

Increasingly, we all live in, or near, cities. A study of accessibility found that "90.7% of people in high-income countries (concentrated in Europe and North America) live within one hour of a city compared to 50.9% of people in low-income countries (concentrated in sub-Saharan Africa)" (Weiss et al, 2018). Accessibility, the ability to get to a city, is not necessarily the same as commuting distance; commuters may live in the city as well as outside it. Table 5.3 shows reported figures for the journey to work in the countries where the cities are located. Based on commuting figures rather than accessibility, people in some low-income countries—Indonesia, India

TABLE 5.3 The cities related to GNI, motor vehicles and national daily one-way commuting distance

City	*Nation*	*GNI*[1]	*Motor Vehicles/capita by nation (2014)*[2]	*One-way journey to work distance in km by nation (2014)*[3]
Higher income			**Average 534**	
Singapore	Singapore	88,155	149	N/A
Dubai	UAE	67,462	313	1.00
New York	United States	63,826	797	8.79
Copenhagen	Denmark	58,662	480	22.67
Amsterdam	Netherlands	57,707	528	3.00
Hamburg	Germany	55,314	572	5.03
Sydney	Australia	48,085	717	4.17
Auckland	New Zealand	40,799	712	N/A
Middle income			**Average 260**	
Kuala Lumpur	Malaysia	27,534	361	7.00
Moscow	Russian Federation	26,157	293	5.67
Buenos Aires	Argentina	21,190	314	N/A
Beijing	China	16,057	83	0.70
São Paulo	Brazil	14,263	249	10.83
Lower income			**Average 33**	
Cairo	Egypt	11,466	45	39.82
Jakarta	Indonesia	11,459	60	6.42
Havana	Cuba	8,621	38	N/A
Ahmadabad	India	6,681	18	5.39
Nairobi	Kenya	4,244	24	7.00
Pyongyang	North Korea	N/A	11	N/A

1 Conceição (2020:343–345).
2 Our World in Data (n.d.a).
3 Nationmaster (n.d.).

and Kenya—are reportedly commuting a shorter distance than people in high-income Denmark and the United States.

Where possible, the subsequent analysis will use commuting data for each city where figures are available, which sadly is not always the case. Table 5.4 combines the national commuting data from Table 5.3 with data for average one-way public transport distance for individual cities where available. Obviously, these are not comparing the same things but both show a lack of relationship between population density, and commute and public transport travel distance. Low-density Copenhagen has a long commute and much higher density. São Paulo has a relatively long average commute and short public transport distance, whereas Dubai has the shortest commute and the second longest public transport travel distance.

TABLE 5.4 The 19 cities, ranked by population density as defined by Demographia (2020:22–92) and national one-way journey to work distance in 2014 and one-way public transport distance

City	*Population density—people per km²*	*One-way journey to work distance in km by nation (2014)*[1]	*One-way travel distance by public transport by city*[2]
Ahmadabad	20,583	5.39	—
Pyongyang	12,651	—	—
Singapore	10,981	—	6.3
Jakarta	9,756	6.42	12.1
Cairo	9,639	39.82	—
São Paulo	7,076	10.83	7.1
Nairobi	7,065	7.00	—
Havana	5,054	—	—
Buenos Aires	5,015	—	9.9
Beijing	4,658	0.70	—
Dubai	4,004	1.00	11.3
Kuala Lumpur	3,831	7.00	7.1
Amsterdam	3,608	3.00	—
Moscow	2,908	5.67	8.5
Auckland	2,548	—	—
Hamburg	2,485	5.03	8.9
Copenhagen	2,187	22.67	—
Sydney	2,123	4.17	8.5
New York	1,700	8.79	8.0

1 Nationmaster (n.d.).
2 Moovit (n.d.).

5.3.1 *The relationship between commuting and total travel*

Darido et al. (2014:147) found that in several Chinese cities, the number of trips per day increased as income increased. The increase in trips suggests that as people get more money, they are able to afford to travel for purposes other than going to work. The same researchers also found the number of households owning private cars increased in all the Chinese cities as income rose (Darido et al., 2014:149).

The effect of wealth has been clearly stated with regard to air travel: the rich fly more than the poor (CAPA, 2014) and the same effect might apply to transport in cities. To explore this in more detail and to put the effect of relative wealth into clearer perspective, Table 5.5 lists the 13 cities (out of the 19 studied in this book) for which data are provided in the Deloitte City Mobility Index (DCMI) (Deloitte, 2021). This provides a breakdown for the modal split of transport for each city, including walking, and is used here as a way of ensuring a common basis for the data. For some cities, the commuting modal split may be different from the DCMI split, which covers the whole day, but the DCMI will be used where possible for consistency and to allow each city's overall transport emissions to be calculated using a wealth-based multiplication factor as explained below. Table 5.5 also shows the gross national income (GNI) for each city (Conceição, 2020:343–345) and the cities are ranked in the table from the wealthiest to the poorest. Some of the city data provided in the DCMI relates to 2020 and some to 2018 and these latter cities are marked with an asterisk.

Table 5.5 shows that there is no clear link between a city's relative wealth and how people travel. The findings are quite surprising. The wealthiest city, Singapore, has the most people using public transport in the higher income group but also twice as many pedestrians as more temperate Copenhagen, while the Danes and the Kenyans are shown to cycle even more than the Dutch. The second poorest

TABLE 5.5 Thirteen of the 19 cities ranked by GNI and showing % modal split of transport based on DCMI

City	*Nation*	*GNI*	*Car*	*Public Transport*	*Bicycle*	*Walk*	*Other*
Higher income							
Singapore	Singapore	88,155	33	53	2	12	-
Dubai	UAE	67,462	83	15	1	1	-
New York*	United States	63,826	57	35	1	6	1
Copenhagen*	Denmark	58,662	26	27	41	6	-
Amsterdam	Netherlands	57,707	42	19	30	5	4
Sydney	Australia	48,085	65	27	1	5	2
Auckland	New Zealand	40,799	81	12	1	5	1
Middle income							
Moscow*	Russia	26,157	19	78	1	1	1
Buenos Aires*	Argentina	21,190	14	75	3	1	4
Beijing*	China	16,057	28	36	9	27	-
São Paulo	Brazil	14,263	36	31	-	32	1
Lower income							
Jakarta	Indonesia	11,459	88	10	0	1	1
Nairobi*	Kenya	4,244	13	46	39	1	1

city, Jakarta, has the highest car use of all the cities, but the people in wealthy Dubai and Auckland are nearly as bad. Moscow and Buenos Aires stand out for their high public transport patronage and low car use.

While wealth may be a poor predictor of the choice of transport modes made by people, it may be a useful determinant of the kind of trips made. If you are poor, you have to get to work if you have a job, but if you have more money, you have the choice to make more non-essential journeys. A detailed breakdown of household travel in New Zealand shows the various kinds of trip people were making in a relatively wealthy country. The division of time spent travelling as interpreted from the New Zealand Household Travel Survey 2011–2014 (Ministry of Transport, 2015:Figure 7.21) is shown in Table 5.6.

TABLE 5.6 Weekly travel of all people in New Zealand by trip length in hours

Purpose of trip	*Hours*	
Shopping/personal business/medical/dental	50	23%
Social visits	41	19%
Accompanying or transport someone	35	16%
Work—main/other job	34	16%
Recreation	25	12%
Work—employer's business	18	8%
Education	13	6%

The commuting journeys of those in the workforce (including walking to and from the bus stop or station) were only 16% of all time spent by New Zealanders in travelling; 84% of all travel was for purposes other than commuting.

Data for England show that only 15% of all trips made per person per year are for commuting and 19% are for shopping (Department for Transport, 2020). These figures from New Zealand and England, one based on time spent travelling and the other on total trips, lend support to the argument that in wealthier countries, people travel more than they do in poorer countries. They spend more of their total time on non-commuting trips and make more non-commuting trips in total.

In an attempt to extrapolate a city's total transport emissions from the calculation of commuting emissions, it is proposed here to add an income-based multiplication factor to reflect the degree of extra travel engendered by prosperity. On the basis that cities are where wealth tends to be concentrated in a nation, it is suggested that for the three categories of cities in Table 5.1, lower income, middle income and higher income, and looking at the relative scale of income in each bracket, the multiplication factors could be 2, 3 and 5, respectively. These multipliers account for trips to school, for shopping and for non-work travel at weekends. The multipliers will tend to underestimate the emissions for public transport because during commuting times, the rush hour, public transport vehicles are likely to be at full capacity and will be more efficient. These multipliers are used in the study of four representative cities in depth in the following sections: two low-income cities, one middle-income and one high-income city.

5.4 Introduction to transport in selected cities

Having looked at the nations in which the cities are located, now we need to try to discover how transport is arranged in the cities themselves. In general, the cities are likely to contain wealthier populations than the surrounding areas, as discussed in Chapter 1.

For some cities data can be found easily, while for others there are fewer published details. In the case of Pyongyang in the North Korea (DPRK), there is little published information available. Is it still possible to draw some conclusions about the city with regard to transport?

5.5 Transport in Pyongyang

One reference states Pyongyang has a population of 2,228,000 (Demographia, 2020:22–92). Another (CIA, 2021) states there were 3.108 million people in 2021. The CIA estimates the national population at 25.8 million and the labour force at 14 million, giving a workforce of just over half the national population. If we assume that half of the city population are working-age adults and that all work, there will be between 1.1 and 1.5 million people going to work each day, depending on which population figure is used. How do they travel? An article with the helpful title "How Do North Koreans Get to Work? A Guide to Transport in the DPRK" (Hill, 2017) provides some of the answers.

5.5.1 Pyongyang's travel modes

According to this article about travel to work, Pyongyang has three public transport systems comprising the Metro, an underground railway with stations 110 metres deep to serve as bomb shelters, plus trolleybuses and trams on the surface. Because oil is scarce, diesel buses have tended to be replaced with electric trolleybuses. All three of Pyongyang's public transport modes are electric; the methods used to generate electricity in the DPRK are discussed in detail later in order to calculate the carbon dioxide emissions.

The surface modes of transport are more popular because they avoid the extra ten-minute travel on the long escalators needed to get down to the deep underground and back up again at the end of your trip. The Metro trains themselves were imported from West Germany in the 1960s. The trolleybuses and trams have a larger network than the underground and are said to run as regularly as underground trains, which run every five minutes at peak times and ten minutes at others. The trolleybuses are mostly locally built while the trams were bought second-hand from Prague.

In addition to these public transport modes, there are at least four taxi companies in Pyongyang, but taxis are very expensive compared to the public transport options. For personal transport, bicycles are used widely, having been illegal until 1992. Locally made bicycles now cover at least 70% of the market. There are also locally made cars, but these are not common.

5.5.2 Estimating travel distance

Is it possible to make a plausible estimate of how many people use the various modes of travel in Pyongyang? Daily ridership on the Metro is estimated to be between 300,000 and 700,000 according to a Swedish travel company (Korea Konsult, n.d.). Ridership must cover total use in the day so half the trips would be into work and the other half would be back home, giving between 150,000 and 350,000 commuters on the Metro.

Using the CIA figure for population and the highest figure for Pyongyang commuters on the Metro means that there must be roughly 1.2 million people using surface transport rather than going underground. We might assume half of them are in trams and trolleybuses with the rest on foot or cycling. Given that the DPRK is not a wealthy country, it seems likely that travel to work will be the main use of transport as money may not be available for more frivolous travel purposes. Looking at the data on the 19 cities in Table 1.3, Pyongyang can be seen as being small in area and high in population density, which implies that commuting distances will be relatively short.

A study of 77 countries in 2013–2014 (Nationmaster, n.d.) showed a range of daily one-way commuting distances between nearly 40 km (Egypt) and 0.7 km (China). For South Korea, the distance is given as 3.0 km, the same as the Netherlands and Argentina. This would be an easy distance to cycle. Since South Korea is the closest country to the DPRK, and since Pyongyang is small and densely populated, this figure is used here.

5.5.3 Estimating energy consumption

We also need to know the likely energy consumption of the transport vehicles.

5.5.3.1 Metro

A study by the Public Transport Users' Association in Australia found that a Melbourne electric commuter train (standing in here for a 1960s Pyongyang Metro train) fully loaded used 0.04–0.18 MJ per passenger-km depending on passenger loading. This is between 0.01 and 0.05 kWh per pass-km (PTUA, 2015a). The lower figure will be used here to represent a fully loaded Pyongyang Metro train.

5.5.3.2 Trams

Melbourne trams in 1991 were estimated to use 12 MJ per vehicle kilometre, or 3.3 kWh (PTUA, 2015b). Trams made by the Czech firm Tatra in the 1970s, which are representative of Pyongyang's trams, had a passenger capacity of around 100 (Strassenbahn-online, 2020). This would give a consumption of 0.03 kWh per pass-km for a full tram.

5.5.3.3 Trolleybuses

The most widely used trolleybus in Pyongyang is the Chollima 091, made in the DPRK (Urban Electric Transit, 2021). It is an articulated bus with a capacity of 170 passengers (China Motor Vehicle Documentation Centre, n.d.). A Polish study found that "the total energy consumption of a standard trolleybus can vary between 1 kWh/km and 4 kWh/km" (Wołek et al., 2021). It will be assumed here that the higher figure will apply to the trolleybuses of Pyongyang. Consumption of a full Chollima 091 will be 0.02 kWh per pass-km.

5.5.3.4 Bicycles

Ritchie (2020) gives a carbon footprint for a bicycle of between 16 and 50 g CO_{2eq} per km depending partly on what food powers the cyclist. A figure of 20 grams will be used here on the assumption that the diet in Pyongyang is low in meat and high-emission processed foods (see Table 3.10).

We now have the data to construct a table. Table 5.7 assumes that all commuters are travelling the same distance and 350,000 of them are using the Metro, travelling 6 km per day (3 km each way). Of the remaining 1.2 million, half are assumed to use non-powered transport (feet and bicycles) and the remaining 600,000 are split equally, so half (300,000) are on trolleybuses and the other half on trams.

TABLE 5.7 Passenger transport modes, passenger numbers and estimated energy consumption

Mode	*Energy per pass-km*	*Total pass-km/day*	*Total energy/day*
Metro	0.01 kWh	2.1 million	21 MWh
Trolleybus	0.02 kWh	1.8 million	36 MWh
Tram	0.03 kWh	1.8 million	54 MWh
Bicycle*	[20 grams CO_2 per pass-km]	1.8 million	[36 tonnes CO_2]
Walking*	Assumed nil	1.8 million	Nil
Daily total			111 MWh
Annual total	*300 days***		*33.3 GWh*

* Although Bicycle and Walking are not calculated on an energy consumption basis, they are included here for the sake of completeness in a table of transport modes.

** Assuming 300 days per year for work-related travel.

5.5.4 Carbon emissions

With regard to emissions, much of the travel to work in Pyongyang is powered by electricity, potentially an ideal situation in regard to reducing carbon emissions. It appears that the annual consumption of electricity for travel to work in Pyongyang might be around 33 GWh.

According to the CIA *World Factbook* (CIA, 2021a), 45% of DPRK electricity is from fossil fuels and 55% from hydro, with annual consumption being 13.89 billion kWh (2016 estimate) or 13,890 GWh. Electricity blackouts have been common, leading to the widespread use of small photovoltaic panels and batteries for domestic use in small appliances and LED lighting. Electricity generation per capita is quoted as around 750 kWh, compared with 9,000 kWh in South Korea (Lankov, 2017).

The CO_2 emission for electricity generation from coal is around 1,000 gCO_{2eq}/kWh (Parliamentary Office of Science and Technology, 2006:2). The same source (2006:3) gives 10–30 gCO_{2eq}/kWh for hydroelectricity. Hydro emissions are somewhat contested due to the methane that may be produced when land is flooded by a new hydro dam. One study suggests a much higher worldwide average carbon footprint of 273 g CO_{2eq}/kWh for hydroelectricity (Scherer and Pfister, 2016).

Using these figures, a 50/50 mix of coal and hydroelectricity would result in an emission of 637 g CO_{2eq}/kWh, using the highest emissions figure for hydroelectricity. The daily emissions from commuting would be 70.7 tonnes. If the emissions for cycling, calculated in Table 5.7 are added, the daily figure is just over 100 tonnes. This gives a total annual figure for carbon dioxide equivalent emissions from transport to work in Pyongyang of 32,010 tonnes, which is around 10.3 kg per person per year for the entire city population.

The national data in Table 5.2 above show an annual per capita emission for land transport in the DPRK of 0.1 tonnes or 100 kg. One reason for the large difference is that category "land transport" includes all vehicles, not only passenger vehicles, and also all freight and other transport, not only passenger transport. In addition, the low emissions for Pyongyang are partly because the city uses electricity to power most of its transport. Even with coal making up half of the electricity generation, electrically powered vehicles are generally far more efficient. According to the US Department of Energy (n.d.), electrically powered vehicles "convert over 77% of the electrical energy from the grid to power at the wheels. Conventional gasoline vehicles only convert about 12%–30% of the energy stored in gasoline to power at the wheels." This applies to vehicles using batteries, while Pyongyang's public transport vehicles take electricity direct from the mains, making them even more efficient.

5.5.5 What if people in Pyongyang drove to work?

If capitalism were to triumph, leading to the complete privatisation of transport, the commuters in Pyongyang would have to buy private cars and drive them to work rather than using electric public transport or bicycles. What would this do to their emissions? Figure 5.1 shows that a medium car has an emission of 192 grams per km. If the people currently using public transport or riding bicycles all drove to work in such cars, they would emit 1,440 tonnes of CO_2 each day, compared with 100 tonnes per day in the current situation. Per capita emissions would rise dramatically in spite of half of electricity in the DPRK being from coal and despite assuming very high hydro-related emissions for the electricity supply in Pyongyang in the calculations.

5.6 Transport in Havana

Havana is the capital and commercial centre of the Republic of Cuba, a Communist country like the DPRK. Havana's population in 2021 was 2.143 million (CIA, 2021b). Transport in Cuba as a whole is quoted as being responsible for about 7% of Cuba's national greenhouse gas emissions compared to about 30% in the wealthy economies of the world (Gutiérrez et al, 2020:9). These are similar figures to those which were calculated for various countries in Table 5.2.

The average salary for a Cuban is said to be between US$25 and US$30 per month, but the state provides subsidised food to every citizen, subsidised or free housing and utilities, free health care and free education. A cup of coffee or an ice cream costs 4 US cents, a sandwich is 8 cents and a pizza 28 cents. A beer costs US$1 and a bottle of rum $2, while a ride in a shared taxi is 40 cents or ten coffees (Isabelle, 2019).

5.6.1 Motor vehicles in Havana

One reason for Cuba's low figure for transport emissions is its low ranking in overall motor vehicle ownership. In 2014, the latest date for which figures are available, Cuba had 38 motor vehicles per capita while the United States had 797 (Our World in Data, n.d.a). The definition of motor vehicles here comprises all powered road vehicles except motorcycles and other two-wheelers.

The low number of motor vehicles is partly due to relative poverty. As we saw in Tables 1.2 and 1.8, Havana's 2.1 million people have a gross national income of around $8,600 each, compared with $88,000 for people in Singapore. There is not much money for buying motor vehicles.

At the end of the 1980s, roughly 85% of Cuba's trade was with the politically sympathetic countries of the Soviet bloc. The collapse of the Soviet Union meant the loss of much of this income and no obvious way to replace it. In response to the loss of income, the government of Cuba had to introduce a range of emergency measures including rationing of food and clothing and introducing ways to reduce reliance on motorised transport (Warren et al, 2015). New vehicles could not be obtained and old ones had to be kept running.

The disappearance of the Soviet trading system deprived Cuba of foreign exchange, reducing import capacity by 75%. These events led to the adoption of emergency measures, with the government tightly rationing scarce resources, including oil, resulting in reductions in the supply of public utilities, scheduled blackouts and reduced industrial production. From 1989 to 1993, fuel imports were cut by 76%, and imports of transport equipment fell by 86%; the government also began to introduce economic reforms to increase hard-currency revenues through foreign investment and restructuring the Cuban economy to become more market based. At the same time as cutting supplies of fuel, vehicles and spare parts, and severely curtailing investments in road, rail, and marine/port infrastructure, the Cuban authorities introduced several innovative behavioural and technological solutions (Enoch et al, 2004; Jaffe and Soligo, 2000; Warren and Enoch, 2006). These included:

- Promoting walking and cycling and the use of animal traction for both agriculture and local transport, especially outside Havana.
- Reorganising bus routes and bus systems in order to maximise efficiency and increase patronage (in Havana).
- Discouraging unnecessary or unwarranted journeys as well as the proactive provision of collective taxis, organised hitch-hiking and an acceptance of higher occupancy in all modes of transport along with fuel rationing.
- Encouraging employers to provide transport to and from work for their employees in the form of "enterprise buses".

Critics have pointed out that Cuba's resulting legacy of outdated motor vehicles means greater localised polluting emissions because the old vehicles lack more modern pollution controls such as catalytic converters and fuel injection (Díaz-Briquets and Pérez-López, 2000:164–202). This is probably true, but these are local emissions. The presence of up-to-date vehicles would do little to reduce carbon emissions, which are of global rather than local impact. The reason that Cuba has very low transport-related carbon emissions compared to wealthier countries is that it has many fewer oil-burning motor vehicles than those wealthy countries. Almost all its powered transport is powered by oil.

5.6.2 Havana public transport

5.6.2.1 Buses

Havana's public transport system is based on conventional diesel buses. Buses are known locally as *guaguas* (pronounced "wawa" after the US manufacturer of Cuba's first buses, Washington, Walton and Company Incorporated, or Wa Wa Co. Inc.) (Whiting, 2019). Cuba is cited as having just over 23,000 buses in total, giving a fleet of 10.9 buses per 1,000 people. Havana, on the other hand, is claimed to have only 0.4 buses per 1,000 people, with a fleet of 928 buses. At the same time, Havana's daily figure for "bus passengers" in 2012 was said to be 860,000 (Warren et al., 2015). It is not clear if this figure is based on total journeys. If it is journeys, there must have been around 430,000 people travelling on Havana's buses each day, assuming each person returned the same day. With a fleet of only 928 buses, each bus would have been carrying around 460 passengers a day.

The figures for bus numbers are fleshed out from an article in Cuba's national online encyclopaedia *EcuRed*, which says that in 2007 Havana bought 650 Chinese buses, 200 from Russia and 100 from Belarus, making a total of 950 (Colaboradores de EcuRed, n.d.a). Of these new buses, 650 are articulated so will hold more passengers. The *EcuRed* article says that the aim of the new bus fleet is to restore bus use to a level of 85% of the historical peak in 1988, when almost 4 million passengers were carried daily. The latter figure suggests that most of the city's population, children as well as adults, used the bus each day.

The website of Havana's public transport system states that the separate components of the bus system in Havana were brought together into one organisation in October 2014, as the Empresa Provincial de Transporte de La Habana, the Provincial Transport Company of Havana (Empresa Provincial de Transporte de La Habana, 2021). This operates two bus networks, the *Red Principal*, which provides a bus equivalent of a Metro on fixed routes, and the *Red Alimentadora*, which provides feeder services to connect to the *Red Principal* (*Red* here has nothing to do with Socialism; it is the Spanish for network). A recent article by a travel company says that the buses run every 10 minutes at peak time and cost US$0.05 per trip (ViaHero, 2020). A recent quote from a blogger confirms that you can get to anywhere in Havana on public transport for the equivalent of US$0.05 (Carter, 2021).

5.6.2.2 Railways

Cuba as a whole ranks number 26 in the world for the length of railway (CIA, 2021c), while with a population of just over 11 million, it is ranked 83rd (CIA, 2021b) by population, suggesting that it has quite a lot of railway track per person. Cuba has at least the potential for much railway use.

The Cuban ministry of transport stated that in 2018 trains carried 6.7 million passengers compared with nearly 11 million passengers in 2004. The government was hoping to increase long-distance passenger numbers by one million in 2019 (Anon, 2019). In 2019 the railways were cited as having 580 locomotives, 20 railcars, 265 passenger coaches and 13,468 freight wagons (Colaboradores de EcuRed, n.d.b). These figures are for the national railway system. Havana also has a modest suburban rail system, but details about it are scarce, largely limited to an entry in Wikipedia (Wikipedia, 2020), which also provides a network map. The Central Station is currently being rebuilt as part of the upgrading of the whole Cuban railway system (Reinaldo and Pérez, 2021). Cuba's railways are entirely diesel powered, apart from the vintage electric Hershey Railway built to serve a sugar plantation (Colaboradores de EcuRed, 2020).

5.6.2.3 Collective taxis

Havana also has shared taxis that travel along a fixed route, more like a bus, and which can be flagged down anywhere along the route. The standard fare is 10 CUP (Cuban Pesos), which is about 50 US cents for any distance along the route (ViaHero, 2020).

5.6.3 *Going to work in Havana*

The labour force in Cuba as a whole in 2017 was estimated at 4.691 million and the July 2021 estimated population was 11,032,343 (CIA, 2021b). This is a workforce of roughly 40%. If Havana's population of 2.14 million follows the national figure, there will be something like 860,000 people going to work each day.

It is probably reasonable to assume that very few people in Havana commute by rail and that people are on foot, on bicycles or using the buses or occasionally collective taxis. Going to work in a shared taxi would cost $4.00 a week, too much to afford every day out of an income of $30 a month. A week's commuting on the bus would be a more reasonable US$0.50. In a description of daily life in Havana based on interviews with a Havana family published in 2018, one family member described taking the bus to work and occasionally a shared taxi if he was late or if the bus was too full (Chicca, 2018:82).

Using only the bus, the cost of commuting would represent around 7% of income. In the United States, a "poverty budget" breaks down into 33% for food, 30% for housing, 27% for all other expenses and 10% for commuting (Bureau of Transportation Statistics, 2012). Given that "Cuba provides their own version of food stamps to every single Cuban citizen, subsidized or free housing and utilities, free health care, and free education" (Isabelle, 2019), the cost of commuting in Havana may be less onerous than in the United States, where fewer services are available to citizens free of charge.

5.6.4 *What mode?*

To begin to calculate the impact of transport, we need to know what percentage of people used the various available modes of transport. Table 5.8 shows the transport modal split, or the percentage usage of various modes of travel, for Havana for 1981, before the collapse of the Soviet Union, and for 1998, during the period of economic shortages. The table is derived from Enoch et al. (2004) using data from Alepuz and Valdés Ríos (2002) plus additional assumptions for 2021.

In 1981 the enterprise bus and the camello did not exist. If they are added to the public bus passengers in 1998, a total of 27% of travellers were using buses in 1998.

In order to have some numbers to work with, a set of estimates for 2021 have been added to the figures from 1981 and 1998 in Table 5.8. These estimated figures assume an increase in bus use compared with 1998 but not yet back to the figure in 1981. They assume some decrease in walking and cycling from the extreme economic situation of 1998; that collective taxis have replaced car use; and that the use of other modes, such as the suburban railway, has declined.

Table 5.4 at the start of this chapter shows that public transport distance is quite similar across a broad range of urban densities while overall commuting distance varies more widely. However, neither a national figure for Cuba nor a specific figure for Havana is provided. In an attempt to find a figure to use for Havana, Table 5.9 shows the one-way commuting distance on public transport in 13 Spanish cities

TABLE 5.8 Modal split of travel in 1981 and 1998

Mode	*% 1981*	*% 1998*	*Assumed % in 2021*
Pedestrian	6.0	43.0	25%
Bicycle*	< 1.0	13.5	10%
Enterprise bus**	-	10.8	5%
Other***	3.5	10.8	5%
Public bus	84.5	9.5	50%
Camello****	-	6.7	-
Car	6.0	5.6	-
Collective taxi	-	-	5%

* Bicycle use was not measured directly; this is an estimate.
** Enterprise buses are provided by employers to transport workers to and from work.
*** Includes ferries, suburban railway, cicliobus, tricycles, motorcycles and other modes.
**** Camellos were high-capacity buses converted from articulated lorries. The name is due to the two humps created by the need for a raised floor at each end over the wheels, with a lowered section in the centre for boarding.

TABLE 5.9 Public transport one-way commuting distance in 13 Spanish cities

City	*Population*[1]	*Public transport one-way commuting distance*[2] *(km)*
Madrid	3,255,944	9.4
Barcelona	1,621,537	7.4
Valencia	814,208	4.8
Sevilla	703,206	4.0
Zaragoza	674,317	3.2
Malaga	568,305	5.6
Mallorca*	401,270	7.9
Bilbao	354,860	6.0
Vigo	297,332	3.2
Granada	234,325	2.8
Tenerife**	222,417	10.2
Burgos	178,966	2.2
Ciudad Real	74,014	2.0
Average		5.4

* Palma.
** Santa Cruz de Tenerife.
1 World Population Review (2021).
2 Moovit (n.d.).

based on data from a company called Moovit (part of Intel), which is a provider of mobility as a service (MaaS) solutions (Moovit, n.d.).

Given that Spain was the colonial power in Cuba at the time that Havana was developed into the capital, it seems appropriate to assume that Havana will be comparable to cities in Spain. Table 5.9 shows little relationship between city size and commuting distance. For the sake of the calculations, the average distance of 5.4 km in the 13 Spanish cities will be used to represent travel distance for commuting in Havana.

5.6.5 Emissions of different transport modes

In Figure 5.1, the carbon emissions of a bus are given as 105 grams per passenger-km (Ritchie, 2020). Would this be an appropriate figure to use for buses in Havana? Tables 5.10 and 5.11 provide data on many types of buses used in the city of São Paulo in Brazil in 2016, ranging from small 20-seaters to double-articulated vehicles (Dallmann, 2019:6–8).

Diesel fuel has an emission of 2.65 kg CO_2/litre (BEIS, 2019). We can see that a 30-seater bus, with 30 seats occupied and no air-conditioning, would have an emission of 26.5 grams of CO_2/passenger-km, while a large "Articulado" with air-conditioning and full to capacity (seated and standing) would have an emission of 12.9 grams of CO_2/passenger-km. A non-air-conditioned "Básico" or basic bus would need to be carrying only 12 passengers to meet the average of 105 grams assumed by the data in Figure 5.1.

TABLE 5.10 Characteristics of bus types deployed in São Paulo municipal public transit bus fleet in 2016

Bus type	*Length in metres*	*Number of seats*	*Total passenger capacity*
Miniônibus	8.4–9.0	20	35
Midiônibus	9.6–11.5	25–33	54–68
Básico	11.5–12.5	35	74
Padron	12.5	32	87
Padron (15m)	15.0	38	110
Articulado	18.3	37	129
Articulado (23m)	23.0	54	174
Biarticulado	≤ 27.0	53	198

TABLE 5.11 Fuel and energy consumption for diesel buses operating in São Paulo

Bus type	*No AC Litres/100 km*	*No AC kWh/100 km*	*With AC Litres/100 km*	*With AC kWh/100 km*
Miniônibus	30	3.0	35	3.5
Midiônibus	40	4.0	47	4.7
Básico	46	4.6	53	5.3
Padron	55	5.5	63	6.3
Padron (15m)	65	6.5	75	7.5
Articulado	71	7.1	80	8.0
Articulado (23m)	75	7.5	85	8.5
Biarticulado	80	8.0	90	9.0

It is likely that rush-hour buses in Havana are full to capacity, which is supported by the resident described earlier who said that he took a shared taxi if the bus was too full (Chicca, 2018:82). Table 5.12 sets out the emissions per passenger-kilometre for each of the buses assuming they are carrying their full capacity of sitting and standing passengers.

TABLE 5.12 Fuel consumption and emissions for diesel buses operating in São Paulo assuming full passenger capacity

Bus type	*Pax*	*No AC L/100 km*	*No AC CO_2/pass-km*	*With AC L/100 km*	*With AC CO_2/pass-km*
Miniônibus	35	30	23 grams	35	27 grams
Midiônibus	68	40	16 grams	47	18 grams
Básico	74	46	16 grams	53	19 grams
Padron	87	55	17 grams	63	19 grams
Padron (15m)	110	65	16 grams	75	18 grams
Articulado	129	71	15 grams	80	16 grams
Articulado (23m)	174	75	11 grams	85	13 grams
Biarticulado	198	80	11 grams	90	12 grams

When full, the medium-size buses all have very similar emissions, the largest types have noticeably lower emissions and the smallest ones have higher emissions.

5.6.6 Havana's transport emissions

These figures can be brought together using some assumptions derived from Tables 5.10–5.12. It is assumed that bicycle emissions are the same as for Pyongyang. Enterprise buses are assumed to be smaller vehicles with an emission of 30 grams CO_2/pass-km and all other public buses are assumed to emit 20 grams per passenger-km. Collective taxis are assumed at 70 grams/pass-km, based on Ritchie's figure of 200 grams for one person in an SUV (Ritchie, 2020). This allows for the fact that the collective taxis will have more than one passenger (Table 5.13).

TABLE 5.13 Transport mode shares and CO_2 emissions for Havana

Mode	*Assumed % for 2021*	*gCO_2/ pass-km*	*Number of people*	*One-way distance travelled (km)*	*CO_2 per day (tonnes)*	*CO_2 per year (tonnes)**
Pedestrian	25%	Nil	215,000	N/A	-	
Bicycle	10%	20	86,000	5.0	17.2	
Enterprise bus	5%	30	43,000	5.4	13.9	
Other**	5%	20	43,000	5.4	9.3	
Public bus	50%	20	430,000	5.4	92.9	
Collective taxi	5%	70	43,000	5.4	0.5	
Total			860,000		133.8	40,140

* Assuming 300 days per year for work-related travel.

** Includes ferries, suburban railway, tricycles, motorcycles and other modes.

The annual transport CO_2 emissions for Havana are 18.7 kg for each member of the city population. This is somewhat higher than Pyongyang's 10.3 kg, probably due to Cuba's total reliance on oil for transport, whereas transport in Pyongyang is mostly electric.

5.7 Transport in Beijing

Moving into the middle-income bracket from Table 5.1, but sticking with Communist cities, how does China's capital compare? Beijing is a huge city compared to Pyongyang or Havana, with a population of 20.897 million in 2021 (CIA, 2021d). Based on the CIA's figures for the national population and workforce, around 55% of people in Beijing can be assumed to be working, around 11.5 million.

After the Chinese Communist Party came to power in 1949, the way that work was organised meant that commuting was more or less non-existent. Many people lived in "urban village-like" work units called *danwei*. Residents could find their daily needs within the *danwei*, which contained schools, shops and hospitals as well as work and housing (Wang and Chai, 2009). In the early 1980s, China had 2.65 million *danwei* (Björklund, 1986) although by that time the 30-year reign of the *danwei* was coming to an end as China began to move to a more market-oriented society (Xiao et al, 2020). A typical *danwei,* the Beijing No. 2 Textile Factory in the Chaoyang District of Beijing, opened in 1955 and provided educational facilities, medical facilities, canteens and other living facilities. The amenities offered were continuously improved to the point that, by the 1970s, in the residential area located on one side of Chaoyang Road, they included an auditorium, a large dining hall, a kindergarten, a primary school, a middle school, a textile industry school, a staff hospital, a post office, a retail store, baby care rooms, bathrooms and a large number of residential buildings. Also in the production area, located on the other side of the road, as well as the large factory buildings, there were welfare facilities including bathrooms, canteens and nurseries (Xiao et al, 2020).

To give an idea of the size of some of these *danwei* developments, Joint Factory 718, also in Chaoyang, was an electronics manufacturing centre built by the German Democratic Republic (the former East Germany). It employed 20,000 workers who enjoyed comfortable housing provided with on-site facilities such as shops, theatres, libraries, hospitals, schools and swimming pools, which offered education, health care and sporting events. This complex is now a centre for avant garde art (Smith, 2019) while Beijing No. 2 Textile Factory is now an industrial heritage theme park (Tang, 2014).

Danwei had a significant effect on transport in Beijing in times past. Wang and Chai (2009) found that residents of *danwei* commuted shorter distances and made more use of walking and cycling. They concluded:

> Danwei were actually an institutional means to link production and consumption in the same place to achieve political and economic objectives. This study shows that as a by-product, Danwei also seems to support the objective of sustainable transportation. Because of its incompatibility with the market-oriented economic reform, the Danwei system has been under reform since the 1980s.

One of the four pillar industries of China's market-oriented reforms in the 1980s was the automotive industry, along with the oil and chemical industry, the machinery and electronics industry and the construction industry (Zheng, 2018:377–384). Zheng's book, originally published in 1993, stated that "an automobile civilization is an important symbol of the modern industrial civilization" (Zheng, 2018:380). The reforms, which aimed to increase the wealth of the Chinese people, had the unfortunate consequence of decreasing sustainability in transport through the simultaneous decline of the residential-industrial *danwei* and the promotion of the private car for transport. The reforms were highly successful in their own terms, meaning that in 2010 China overtook the United States to become the world's largest car manufacturer (PWC, 2018). This growth in car production had an obvious downside: from 2000 to 2009, Beijing's commuting-related CO_2 emissions were estimated to have increased from 1.13 million tons to 4.04 million tons, with the emissions share of private cars rising from around two-thirds to over 90 percent (Wang et at, 2014).

5.7.1 How do people travel to work in modern Beijing?

Beijing's overall transport modal split has been broken down as 36% public transport (metro, bus and commuter train), 28% private car, 27% walking and 9% bicycle. In spite of these figures with their emphasis on walking, cycling and public transport, Beijing has the largest ownership of private cars in China and a rush-hour commuting time of nearly 53 minutes (Deloitte, 2018a). Total public transport passengers in 2019 were just over 7 billion or nearly 19.5 million per day. In 1978 the large majority of public transport passengers were on buses, but by 2019 rail had overtaken buses as the majority carrier. In 2013 and 2014, public transport passenger numbers peaked at just over 8 billion per year (Beijing Municipal Bureau of Statistics, 2020), suggesting that the promotion of car ownership has succeeded.

5.7.2 Public transport in Beijing

5.7.2.1 The Beijing Subway

As Beijing has grown, its underground railway system, the Beijing Subway, has expanded rapidly. Between 2009 and 2015, the length of track of the Beijing Subway nearly doubled, from 228 km to 554 km (Jiang and Levinson, 2017). According to a travel website, the length in 2020 was 1,117 km with 23 lines, making it the second largest underground in the world after the Shanghai Metro (Song, 2021). Yang et al. (2015) found that the construction of new lines for the Beijing Subway did not result in any measurable decrease in traffic congestion, but each new line reduced daily bus ridership by 452,400. Ridership on the Beijing Subway is said to vary between 9.8 and 11.6 million per day (Song, 2021). An average daily ridership of 10 million per day was quoted in 2014 (Li et al, 2014). These figures fit with official statistics, which show "rail passengers" in 2019 were 3,962,380,000 or 10.9 million per day (Beijing Municipal Bureau of Statistics, 2020), forming 56% of total public transport usage.

5.7.2.2 Buses

In 2019 Beijing had 23,010 buses in total, mostly diesel buses but also some electric trolleybuses. Bus passengers were 3,133,660,000 per year or 8.5 million per day (Beijing Municipal Bureau of Statistics, 2020), making bus passengers 44% of the total of public transport users. Bus trips were cited as forming 29% of all Beijing public transport trips in 2011 (Zhou et al, 2017). From the increase in bus passengers between 2011 and 2019, it does not appear that the construction of new Beijing Subway lines has decreased bus usage.

- Diesel buses

Beijing's buses are relatively modern. A listing of the city's buses shows that all those in service were built in the 21st century (Buspictures, 2021). Diesel buses are increasingly seen as a source of pollution and a few years ago they were reported as being replaced with electric buses on some routes (Chinabuses, 2014), with 60% of the Beijing bus fleet planned to be fully electric by 2020 (Xinhua, 2017).

- Trolleybuses

The intended introduction of electric buses does not seem to have taken place as rapidly as hoped since in 2020 the Beijing public transport system was reported to have only 2,122 "new energy vehicles", of which 1,276 were dual-source trolleybuses (Ren Shan and Zhu, 2020) that can operate for some distance on battery power. The category of new energy vehicles covers hybrid vehicles as well as purely electric ones (Yuan et al, 2015). It is assumed here that the 2,122 new energy vehicles in the bus fleet are all trolleybuses and that they represent 9% of the total bus fleet.

5.7.2.3 Bicycles

A recent estimate says there are 4 million electric bicycles (E-bikes) in Beijing, with an average annual growth of 800,000 E-bikes (Yu et al, 2019). Given this figure, it will be assumed here that the commuters who are cycling are using E-bikes, which are said to have emissions similar to those of a normal bicycle (ECF, 2011:11). The modal split of 9% cycling supports this argument since 9% of the workforce would be around 1 million people, while there are 4 million electric bikes.

5.7.3 The journey to work

If Beijing's workforce follows the national figures, 55% of the population of 20.897 million (CIA, 2021d) are likely to be going to work, giving 11.5 million daily commuters. If they follow the modal split quoted above, the usage figures for the available modes are shown in Table 5.14.

TABLE 5.14 Split of transport modes between workers

Mode	*%*	*Number of passengers*
All commuters	100%	11,500,000
public transport	36%	4,140,000
private car	28%	3,220,000
Walking	27%	3,105,000
Bicycle	9%	1,035,000

In Beijing the category public transport comprises the Beijing Subway, buses and trolleybuses. At first sight, the estimate of around 4 million public transport users does not seem to fit with the 10 million people riding the Subway each day; however, not all travel is for the purpose of going to and from work as discussed in the introduction to this chapter. In addition, each commuter will make at least two trips (to work and back home) meaning that one person will count as two riders.

TABLE 5.15 Split of public transport modes between workers

Mode	*%*	*Number of passengers*
Public transport total	100%	4,140,000
Subway	56%	2,318,400
Bus	44%	1,821,600
Diesel bus	91%	1,657,656
Trolleybus	9%	163,944

5.7.4 Commuting distance and emissions

The average commuting distance in Beijing in 2018 was reported to be about 13 km (Wang et al, 2020) so that figure is used here for motorised forms of transport. A study of cyclists who commuted by bicycle in the United States found that cyclists had a one-way commuting distance of 10.7 km, half the distance for all modes, which was just under 20 km (Antonakos, 1993:58). Given the high availability of electric bicycles, it will be assumed here that Beijing cyclists travel rather more than half the average one-way commuting distance of 13 km and a figure of 8 km will be used. E-bikes are reported to use 10–20 Wh per mile (Anderson, 2019) or roughly 6–12 Wh per km. This is similar to the figure per passenger-km for a full Subway train.

The carbon dioxide equivalent intensity of electricity in China was estimated to vary between 861 and 821 g/kWh in 2020 (Li et al, 2017). An average of 850 g/kWh (850 kg/MWh) will be used here. This information has been used to generate Table 5.16.

In 2009 "Transportation, storage, and postal system" emissions for Beijing were 4,659,300 tonnes, which is 12,765 tonnes per day (Zhang et al, 2013). This is carbon emissions, and the figures must be multiplied by 3.67 to convert to carbon dioxide, giving a daily emission in 2009 of 47,000 tonnes of CO_2. This figure (which is somewhat out of date) can be compared with the 17,495 tonnes for commuting emissions.

The annual commuting CO_2 emissions for Havana and Pyongyang were calculated earlier as 18.7 and 10.3 kg, respectively, for each member of the city population. In Beijing, a much larger city the equivalent figure is 251.2 kg as shown in Table 5.17, which compares the three socialist cities.

Table 5.17 shows that the one-way commuting distance in Beijing is considerably greater than that in the other two cities, but it is only three to five times greater, whereas the emissions per person are up to 25 times greater. The reason lies not in Beijing's size and population but in its workers' choice of modes. The 28% of Beijing workers who commute by car emit ten times the carbon dioxide of the remaining 72%, even though China's relatively high-carbon electricity powers the Beijing Subway and trolleybuses.

5.7.5 The effect of electric cars in Beijing

What would be the effect if electricity powered the cars? China is making a great effort to become a major producer, and user, of electric cars. A recent mandate requires electric vehicles (EVs) to be 40% of all car sales by 2030 (Stauffer, 2020). How much difference might a move to EVs make to Beijing's transport emissions? A listing of all battery-only EVs shows consumption from around 15 kWh/100 km to around 25 kWh/100 km (EV Database, 2021). Assuming a mid-range figure of 20 kWh/100 km, the emissions of Beijing's current car commuters can be compared with the same number of people commuting the same distance in EVs powered by China's current mix of electricity generation (Table 5.18). The effect on the city's emissions is quite minimal.

The reasons for this are two-fold. The first is that electricity generation in China is highly dependent on coal. In 2017 coal made up 72% of electricity generation, a figure which is expected to decrease over time but is still expected to be 47% by 2040 (EIA, 2017). This means that electricity in China will have a high-carbon-related emission for many years in the future. The other is related to the efficiency of EVs. Compared to petrol-powered cars, EVs use significantly less energy; a litre of

TABLE 5.16 Beijing commuting CO_2 emissions

Mode	*Mod-al split %*	*CO_2/ pass-km*	*kWh per pass-km**	*Number of people*	*One-way (km)*	*Electricity usage (MWh)*	*CO_2 kg/MW-h*	*CO_2 per day (tonnes)*	*CO_2 per year (tonnes)***
Private car	28%	190 g	—	3,220,000	13	—	—	15,907	
Pedestrian	27%	Nil	—	3,105,000	n/a	—	—	—	
Subway	20%	—	0.01	2,318,400	13	603	850	513	
Diesel bus	14%	20g	—	1,657,656	13	—	—	862	
Bicycle	9%	—	0.01	1,035,000	8	166	850	141	
Trolleybus	1%	—	0.02	163,944	13	85	850	72	
Total		—	—	11,500,000				17,495	5,249 million

* For modes powered by electricity.

** Assuming 300 days per year for work-related travel.

TABLE 5.17 Comparison between the three socialist cities

City	*Population*	*One-way commute (km)*	*CO_2 per person (kg)*
Pyongyang	3,108,000	3.0	10.3
Havana	2,143,000	5.4	18.7
Beijing	20,897,000	13.0	251.2

TABLE 5.18 The effect of changing to EVs for commuting, all other conditions unchanged

Mode	*Modal split %*	*CO_2/ pass-km*	*kWh per pass-km*	*Number of people*	*One-way*	*Electricity usage (MWh)*	*CO_2 kg/ MWh*	*CO_2 per day (tonnes)*
Private car	28%	190g	-	3,220,000	13km	-	-	15,907
Private EV	28%	-	0.20	3,220,000	13km	16,744	850	14,232

petrol contains very roughly 10 kWh (Appropedia, 2021), meaning that an average EV that achieves a performance of 20 kWh/100km is using the energy equivalent of 2 litres of petrol per 100 km, equivalent to 140 miles per gallon (Imperial), three times less than even a very efficient petrol car.

However, comparing Table 5.18 with Table 5.16 shows that even such efficient vehicles are using 20 times the energy per passenger-kilometre of a Beijing Subway train and have nearly ten times the emissions of a diesel bus in the rush hour. The only way for EVs to make any sense in Beijing is if the electricity system were to be radically decarbonised, but even if that were to occur, travelling by public transport would use far less energy overall.

5.7 Transport in New York

We have looked at three cities so far, two relatively small and one huge, but all three governed by a command economy. Two of the cities are in the low-income category and one, Beijing, is a middle-income city. Beijing's per capita commuting emissions are 25 times those of Pyongyang and, unsurprisingly, its car ownership is far greater. In comparison, the United States as a whole is very committed to the private car as the means for getting to work, with nearly 85% of commuters using a "car, truck or van" and only 5% using "public transportation" (Burrows et al, 2021). How does New York, a wealthy city that represents the pinnacle of the democratic capitalist dream, compare with these poorer and ideologically very different communist cities?

5.7.1 Commuters in New York

The CIA (2021e) gives a population of 18.823 million for New York-Newark, while the United States Census Bureau (n.d.a) gives the July 2019 population figures as 8,336,817 for New York city. Table 5.19 shows the transport modal split according to the Deloitte City Mobility Index (DCMI) (Deloitte, 2018b).

TABLE 5.19 Breakdown of travel for the higher income cities

City	*Nation*	*Car*	*Public Transport*	*Bicycle*	*Walk*	*Other*
Higher income						
Singapore	Singapore	33	53	2	12	—
Dubai	UAE	83	15	1	1	—
New York	United States	57	35	1	6	1
Copenhagen	Denmark	26	27	41	6	—
Amsterdam	Netherlands	42	19	30	5	4
Sydney	Australia	65	27	1	5	2
Auckland	New Zealand	81	12	1	5	1

Table 5.19 shows that New York is well above the US average of 5% for public transport use. However, the DCMI data for New York in Table 5.19 refer to the entire NY-NJ-CT region with a population of nearly 19 million. New York city is comprised of the boroughs ofThe Bronx, Brooklyn, Manhattan, Queens and Staten Island (WorldAtlas, 2021). Commuting data for New York city in 2017 (US Census Bureau, n.d.b) give a different split as shown in Table 5.20, with even greater public transport use compared with the wider region.

TABLE 5.20 Transport modal split for journey to workplace in 2017

Mode	*Number*	*Breakdown of public transport users*	*Percentage of total excluding home workers*	*Percentage of public transport users*
Total going to workplace	*4,808,014*			
(excluding home workers)	4,634,710		100%	
Car/truck/van	1,296,486		28%	
Public transport (excluding cab)	2,799,201		60%	100%
Bus		552,879		20%
Streetcar*		9,418		Negligible*
Subway or elevated		1,862,009		67%
Railroad		354,091		13%
Ferryboat		20,804		1%
Bicycle	51,838		1%	
Walk	410,888		9%	
Taxicab/motorcycle/other	76,297		2%	
Worked at home 4% of total	173,304			

* In spite of what the US Census Bureau states, there appear to be no streetcars (trams) in New York city, although there is an aerial tramway (cable car) across the East River (RIOC, n.d.). This category may refer to the AirTrain, a light rail system connecting JFK Airport to the New York Subway (JFK, 2021).

The increase to 60% of people using public transport in New York city, compared with 35% in the DCMI data for the wider region, is striking. The figure is backed by Burrows et al. (2021) who show that 2,242,092 commuters used public transport in 2019, comprising 55.6% of all New York city workers, compared to 5% for the United States as a whole. Two-thirds of these public transport users are on the New York Subway, suggesting that where this option exists (i.e. in the city rather than the wider region) people will use it.

New York city is leading the way in the United States for public transport use, so how does it fare in terms of energy and environmental impact?

5.7.2 New York modes of transport

5.7.2.1 Private car, truck and van

In the United States, "light trucks" comprised 69% of car sales in 2019 (Muller, 2019). The definition of the category of vehicles known as light-duty trucks includes not only pick-up trucks but also vans, minivans and SUVs (National Archives, 2021). These kinds of vehicles are very popular, but their fuel efficiency is poor compared to conventional cars. The average American car gets 24.2 US mpg (9.7 litres/100 km), which itself is poor compared to the EU average of 5.3 litres/100 km in 2016 (calculated from Fontaras et al., 2017 and University of Exeter, n.d.) while the average American light truck/van achieves only 17.5 US mpg (13.4 litres/100 km) (US Department of Energy, 2020). Assuming that inefficient light trucks form two-thirds of the light vehicle fleet, the average fuel consumption for New York's private vehicles will be around 12 litres/100 km.

5.7.2.2 Buses

The buses of New York are operated by the Metropolitan Transit Authority, North America's largest transportation network. It serves 15.3 million people across New York City, Long Island, southeastern New York State and Connecticut and comprises the nation's largest bus fleet (MTA, n.d.). Like many urban bus networks, the New York city bus system is complex, with a number of different kinds of buses, "Local buses, Express buses, Select Bus Service (SBS) buses and Limited buses" (MTA, 2021), and several ways of paying for a ride, which differ between the bus types, all of which no doubt becomes familiar to regular bus users.

5.7.2.3 The Subway

The New York Subway opened in 1904. It now has 472 stations, about 60% of them underground and a total length of 665 miles (1,070 km), served by 6,684 subway cars (MTA, 2020a). Average daily Subway use for 2019 on a weekday was 5,493,875, with a similar number for the two days of the weekend combined 5,494,195 (MTA, 2020b).

5.7.2.4 Suburban trains

Two commuter rail systems serve New York city, the Long Island Rail Road (LIRR), which operates into Penn Station and is the largest commuter railroad in the United States, and the Metro-North Railroad, the second busiest commuter railroad in the United States. Metro-North was formed in 1982 out of the New York Central, New Haven and Erie railroads. It operates from Grand Central Terminal (NYDOT, 2021).

5.7.2.5 Ferries

Including the Coney Island ferry due to start in 2021, New York has eight ferry services offered by one company (NYC Ferry, 2021) and several more from other companies, as well as the Staten Island Ferry, operated free of charge by the city of New York, which operates 24 hours a day and whose vessels include the world's largest passenger carriers (Staten Island Ferries, n.d.).

5.7.3 The journey to work in New York

Questions about commuting are asked by the US Census Bureau (2021) in the annual American Community Survey, but the questions are about mode and time, not about distance. Without knowing the average speed of modes, it is not possible to work out distance. Table 5.4 showed the commuting distance for public transport users in New York to be 8.0 km each way, which is similar to many other cities in Table 5.4, so that figure will be used here for all New York commuters including cyclists.

In 2015 the emissions factor for electricity in New York city was 257 $kgCO_2$/MWh (City of New York, 2017:49). This factor is used to calculate emissions for electrically powered transport modes in Table 5.21. It is assumed that people work for 250 days per year (50 weeks, 5 days a week) less than in the less wealthy communist regimes.

TABLE 5.21 Emissions related to commuting in New York city

Mode	*Number of people*	*CO_2/ pass-km*	*kWh per pass-km*	*One-way*	*Electricity usage per day*	*CO_2 per day (tonnes)*	*CO_2 per year (tonnes)*
Private car	1,296,486	200g	-	8.0km	-	4149	1,037,250
Bus	552,879	20g	-	8.0km	-	177	44,250
Streetcar	9,418	-	0.02	8.0km	3MWh	0.8	200
Subway	1,862,009	-	0.01	8.0km	298MWh	77	19,250
Railroad	354,091	-	0.02	8.0km	113MWh	29	7,250
Ferryboat*	20,804	20g	-	8.0km	-	7	1,750
Bicycle	51,838	-	0.01	8.0km	8MWh	2	500
Walk	410,888	-	-	-	-	-	-
Other**	76,297	200g	-	8.0km	-	244	61,000
Total						4,686	1,171,500

* From Figure 5.1 for foot passengers.

** Assumed as for car/truck/van.

The population of New York is 8,336,817, giving an annual emission of 141 kgCO_2 per person, which is less than Beijing's 251 kg per person. The emissions per day for the total population are closer, at 840 g per person for Beijing and 560 g for New York, because it has been assumed that New Yorkers commute 250 days a year, compared with 300 days in Beijing.

New York and Beijing have a similar around 30% of residents using private vehicles to commute. The main reason for New York's lower per capita emissions is how its electricity is generated. New York state's electricity generating mix is roughly natural gas 40%, nuclear 30% and hydro 25%. The United States as a whole is similarly 40% natural gas 40% with coal 20% and nuclear 20%, with the remainder being mostly wind 8.5%, hydro 7% and solar 2% (US Department of Energy, n.d.). In comparison, in China coal made up 72% of electricity generation in 2017, a figure which is expected to decrease over time but is still expected to be 47% (more than twice the current US figure) by 2040 (EIA, 2017). The way electricity is generated makes a big difference. It was shown above that in Beijing a move to electric cars would make little impact on the city's commuting emissions. In New York however, moving to EVs would cut the emissions of those who commute in private vehicles from 1,037,250 tonnes to 285,227 tonnes.

5.8 Last words

The four case studies demonstrate the problem of finding data to make comparisons of environmental impact in terms of greenhouse gas emissions. Probing the data for Beijing further showed the local transportation authority, the city council and the national government have different information on Beijing and the Beijing region. Contact via e-mail and telephone failed to resolve these issues. What the case studies do show is that comparisons have to be made through best estimates using the available data. What can be drawn from this is that the mode of travel is more important than the size of city, and length of commute and how electricity is generated is critical. That said, cities that encourage walking and cycling to work will have lower emissions. The other advantage of these travel modes is that those doing them are aware of the energy they are using, unlike those sitting in cars or public transport.

References

Alepuz, M. and Valdés Ríos, H. (2002). The urban transportation in time of crisis: The case of Havana. *Proceedings of CODATU X*, Lome, Togo, 12–15 November 2002.

Anderson, S. (2019). Understanding E-Bike Power, Range, and Energy. *Cycle Volta.*, 22 August 2021.

Anon (2019). CUBA: Gov't seeks to revive once-great rail network. *The Gleaner*. Jamaica Gleaner (jamaica-gleaner.com), 15 June 2021.

Antonakos, C.L. (1993). Environmental and Travel Preferences of Cyclists. PhD in Urban, Technological and Environmental Planning, The University of Michigan.

Appropedia (2021). Energy content of fuels. Appropedia, 29 August 2021.

Beijing Municipal Bureau of Statistics (2020). 8–2 Urban Public Transport. *Beijing Statistical Yearbook 2020.* 6 August 2021.
BEIS (2019). *UK Government GHG Conversion Factors for Company Reporting.* London: Department for Business, Energy and Industrial Strategy.
Björklund, E. (1986). The *Danwei*: sociospatial characteristics of work units in China's urban society. *Economic Geography 62*(1), 19–29.
Bureau of Transportation Statistics (2012). Poverty Budget. US Department of Transportation Bureau of Transportation Statistics. Retrieved from Commuting Expenses: Disparity for the Working Poor | Bureau of Transportation Statistics (bts.gov), 28 June 2021.
Burrows, M., Burd, C. and McKenzie, B. (2021). Commuting by Public Transportation in the United States: 2019. *American Community Survey Reports, ACS-48.* Washington DC. U.S. Census Bureau.
Buspictures (2021). Vehicles List: Beijing, Public Transport Holdings. Retrieved from Beijing, Public Transport Holdings—Vehicles List—BUSPICTURES (busphoto.eu), 26 August 2021.
CAPA (2014). Air travel rises with a country's wealth. Law of nature, or can government policy make a difference? *Centre for Aviation.* Retrieved from Air travel rises with a country's wealth. Law of nature, or can government policy make a difference? | CAPA (centreforaviation.com), 6 August 2021.
Carter, S. (2021). Cuban Transportation Guide—Cuban Transport—buses, trains, taxis and colectivos in Cuba. *A Social Nomad.* 28 June 2021.
Chinabuses (2014). Beijing Electric Bus Promotion Application Action Plan Released. Retrieved from Beijing Electric Bus Promotion Application Action Plan Released - news-www.chinabuses.org, 26 August 2021.
CIA (Central Intelligence Agency) (2021a). Korea, North. *The World Factbook.*, 11 May 2021.
CIA (2021b). Cuba. *The World Factbook.*, 31 May 2021.
CIA (2021c). Country comparisons—Railways. *The World Factbook.*, 31 May 2021.
CIA (2021d). China. *The World Factbook.*, 19 July 2021.
CIA (2021e). United States. *The World Factbook.* Retrieved from www.cia.gov/the-world-factbook/countries/united-states/, 30 August 2021.
Chicca, F. (2018). Silvio and Veronica: Watching television in the street: Cuba. in Chicca, F., Vale, B. and Vale, R. (eds.), *Everyday Lifestyles and Sustainability – The Environmental Impact of Doing the Same Things Differently.* Abingdon and New York: Earthscan, 81–84.
China Motor Vehicle Documentation Centre (n.d.). *Chollima 091.* Retrieved from CHOLLIMA 091 | chinesecars, 19 July 2021.
City of New York (2017). *Inventory of New York City's Greenhouse Gas Emissions.* New York: Mayor's Office of Sustainability.
Colaboradores de EcuRed (n.d.a). Guaguas en La Habana. *EcuRed.* Retrieved from www.ecured.cu/index.php?title=Guaguas_en_La_Habana&oldid=3889767, 14 June 2021.
Colaboradores de EcuRed (n.d.b) Ferrocarriles en Cuba. *EcuRed.* 15 June 2021.
Colaboradores de EcuRed (2020), Tren de Hershey. *EcuRed.* Retrieved from at www.ecured.cu/index.php?title=Tren_de_Hershey&oldid=3631606, 13 July 2021.
Conceição, P. (2020). *Human Development Report 2020.* New York: United Nations Development Programme (UNDP).
Dallmann, T. (2019). *Climate and Air Pollutant Emissions Benefits of Bus Technology Options in São Paulo.* White Paper. February. Washington DC: International Council on Clean Transportation.
Darido, G., Torres-Montoya, M. and Mehndiratta, S. (2014). Urban transport and CO_2 emissions: some evidence from Chinese cities. *WIREs Energy Environment 3*(2), 122–155.

Deloitte (2018a). *Deloitte City Mobility Index: Beijing.* London: Deloitte.

Deloitte (2018b). *Deloitte City Mobility Index: New York.* Retrieved from New York_GlobalCityMobility (deloitte.com), 30 August 2021.

Deloitte (2021). *The 2020 Deloitte City Mobility Index.* Retrieved from The Deloitte City Mobility Index | Deloitte Insights Middle East, 7 August 2021

Demographia (2020). *Demographia World Urban Areas:* 16th Annual Edition. Retrieved from www.demographia.com/db-worldua.pdf, 29 March 2021.

Department for Transport (2020). Average number of trips (trip rates) by age, gender and purpose: England, 2019. *National Travel Survey Table NTS0611.* Retrieved from https://assets.publishing.service.gov.uk/government/uploads/system/uploads/attachment_data/file/906276/national-travel-survey-2019.pdf, 19 August 2021.

Díaz-Briquets, S. and Pérez-López, J. (2000). *Conquering Nature: The Environmental Legacy of Socialism in Cuba.* Pittsburgh: University of Pittsburgh Press.

ECF (2011). *Cycle more Often 2 Cool Down the Planet! Quantifying CO_2 savings of Cycling.* November. Brussels, European Cyclists' Federation ASBL.

EIA (2017). Chinese coal-fired electricity generation expected to flatten as mix shifts to renewables. *Today in Energy.* Washington, US Energy Information Administration, 27 September. Retrieved from Chinese coal-fired electricity generation expected to flatten as mix shifts to renewables - Today in Energy - U.S. Energy Information Administration (EIA), 23 August 2021.

Empresa Provincial de Transporte de La Habana (2021). Nuestra Empresa. Retrieved from Nuestra Empresa | Empresa Provincial de Transporte de La Habana (transnet.cu), 27 June 2021.

Enoch, M., Warren, J., Valdes Rios, H. and Henríquez Menoyo, E. (2004). The Effect of economic restrictions on transport practices in Cuba. *Transport Policy 11*(1), 67–76.

EV Database (2021). Most efficient electric vehicle. Retrieved from Most efficient electric cars - EV Database (ev-database.org), 23 August 2021.

Fontaras, G., Zacharof, N-G. and Ciuffo, B. (2017). Fuel consumption and CO_2 emissions from passenger cars in Europe—Laboratory versus real-world emissions. *Progress in Energy and Combustion Science 60*, 97–131.

Gutiérrez, A.S., Eras, J.J.C., Rueda-Bayona, J.G., Hens, L. and Vandecasteele, C. (2020). Assessing the biofuel—transport nexus. The case of the sugar industry in Cuba. *Journal of Cleaner Production 267*, 18.

Hill, M. (2017). How do North Koreans get to work? A guide to transport in the DPRK. Retrieved from How do North Koreans get to work? A guide to transport in the DPRK—City Monitor, 27 April 2021.

Isabelle, G. (2019). 7 Surprising Facts About Everyday Life in Cuba from Baracoa to Havana. *Dominican Abroad*, 24 July. Retrieved from 7 Surprising Facts About Everyday Life in Cuba from Baracoa to Havana (dominicanabroad.com), 28 June 2021.

JFK (2021). JFK Airtrain. Retrieved from https://jfkairport.com/to-from-airport/air-train, 24 September 2021.

Jaffe, A.M. and Soligo, R. (2000). Energy in Cuba. Report to the J A Baker Institute for Public Policy, Rice University. Retrieved from Estimating Per Capita Petroleum Use:The Cuban Case (rice.edu), 13 July 2021.

Jiang, H. and Levinson, D. (2017). Accessibility and the evaluation of investments on the Beijing subway. *Journal of Transport and Land Use 10*(1), 395–408.

Korea Konsult (n.d.). *Pyongyang Metro—The Deepest Metro in the World.* Retrieved from Pyongyang Metro, North Korea (koreakonsult.net), 27 April 2021.

Lankov, A. (2017). *How North Korea's electricity supply became one of the world's worst.* Retrieved from How North Korea's electricity supply became one of the world's worst | NK News, 27 April 2021.

Li, D., Yin, H. and Zhou, W. (2014). Measuring passenger crowd in subway network: Beijing experience. *Transportation Research Board 93rd Annual Meeting* Washington DC. 12–16 January.

Li, X., Chalvatzis, K. and Pappas, D. (2017). China's electricity emission intensity in 2020—an analysis at provincial level. Paper presented at 9th International Conference on Applied Energy, ICAE2017, 21-24 August 2017, Cardiff, UK. *Energy Procedia 142*, 2779–2785.

MTA (2020a). Subway and bus facts 2019: Transit nerds, this is for you. New York, Metropolitan Transit Authority. Retrieved from https://new.mta.info/agency/new-york-city-transit/subway-bus-facts-2019, 6 September 2021.

MTA (2020b). Subway and bus ridership for 2019. New York, Metropolitan Transit Authority. Retrieved from https://new.mta.info/agency/new-york-city-transit/subway-bus-ridership-2019, 6 September 2021.

MTA (2021). Riding the bus. Retrieved from https://new.mta.info/guides/riding-the-bus, 15 September 2021.

MTA (n.d.). About us. Retrieved from https://new.mta.info/about-us, 15 September 2021.

Ministry of Transport (2015). *Comparing Travel Modes New Zealand Household Travel Survey 2011–2014* March. Wellington: Ministry of Transport/Te Manatū Waka.

Moovit (n.d.). Distance traveled by public transit—How far do people usually commute each way with public transit? *Moovit*. Retrieved from People commute every day by public transit | Moovit Public Transit Index (moovitapp.com), 1 July 2021.

Muller, D. (2019). Light trucks are now a record 69 percent of the U.S. market. *Autoweek*. 7 January. Retrieved from Light trucks are now a record 69 percent of the U.S. market (autoweek.com), 2 September 2021.

National Archives (2021). *Code of Federal Regulations Title 40, Chapter I, Subchapter C, Part 86, Subpart A, § 86.082-2*. 30 August. Retrieved from eCFR:: 40 CFR 86.082-2—Definitions. (federalregister.gov), 2 September 2021.

Nationmaster (n.d.). *Transport > Commute > Distance: Countries Compared*. Retrieved from Countries Compared by Transport > Commute > Distance. International Statistics at NationMaster.com, 28 April 2021.

NYC Ferry (2021). Routes and Schedules: All NYC Ferry Routes. Retrieved from https://ferry.nyc/routes-and-schedules/, 10 September 2021.

NYDOT (2021). *Passenger Rail Service in New York State*. New York, Department of Transportation. Retrieved from www.dot.ny.gov/divisions/operating/opdm/passenger-rail/passenger-rail-service/commuter-rail-service, 6 September 2021.

Our World in Data (n.d.a). Motor vehicles per 1000 inhabitants vs GDP per capita, 2014. *Our World in Data*. Retrieved from Motor vehicles per 1000 inhabitants vs GDP per capita, 2014 (ourworldindata.org), 27 August 2021.

Our World in Data (n.d.b). Total greenhouse gas emissions, excluding land use and forestry. *Our World in Data*. Retrieved from Total greenhouse gas emissions, excluding land use and forestry (ourworldindata.org), 10 June 2021

Parliamentary Office of Science and Technology (2006). Carbon footprint of electricity generation. *Postnote* Number 268, October. London: Parliamentary Office of Science and Technology.

PWC (2018). *The Opening-up of Chinese Automotive Industry and its Impact*. Retrieved from chinese-automotive-industry-opening-up-impact.pdf (pwccn.com), 2 August 2021.

PTUA (2015a). Myth: Public transport doesn't really save energy. Retrieved from Myth: Public transport doesn't really save energy | Public Transport Users Association (Victoria, Australia) (ptua.org.au), 28 April 2021.

PTUA (2015b). Myth: Trams have more greenhouse emissions than cars. Retrieved from Myth: Trams have more greenhouse emissions than cars | Public Transport Users Association (Victoria, Australia) (ptua.org.au), 28 April 2021.

Reinaldo, O. and Pérez, I. (2021). Estación Central de Ferrocarriles: ¿Ya viene el tren? *Cubadebate.* Retrieved from Estación Central de Ferrocarriles: ¿Ya viene el tren? | Cubadebate, 25 June 2021.

Ren Shan, Z. and Zhu, S. (2020). Beijing multi-tso and take the pm2.5 micrograms one microgram down keying. *Beijing Daily* 15 January. Retrieved from Beijing multi-tso and take the governance PM2.5 micrograms one microgram down key - Xinhua (xinhuanet.com), 6 August 2021.

RIOC (n.d.). *Tram.* New York, Roosevelt Island Operating Corporation. Retrieved from Tram | Roosevelt Island Operating Corporation of the State of New York (ny.gov), 2 September 2021.

Ritchie, H. (2020). Which form of transport has the smallest carbon footprint? *Our World in Data.* Retrieved from Which form of transport has the smallest carbon footprint? - Our World in Data, 21 June 2021.

Ritchie, H. and Roser, M. (2020). Emissions by Sector. *Our World in Data.* Retrieved from Emissions by sector - Our World in Data, 10 June 2021.

Scherer, L. and Pfister, S. (2016). Hydropower's biogenic carbon footprint. *PLoS One 11*(9), 11.

Smith, T. (2019). A brief history of Beijing's 798 art district. *Culture Trip.* Retrieved from A Brief History of Beijing's 798 Art District (theculturetrip.com), 26 July 2021.

Song, C. (2021). Beijing Subway. *China Highlights.* Retrieved from Beijing Subway: tourist information, timetable, prices (chinahighlights.com), 4 August 2021.

Staten Island (n.d.). Staten Island Ferries—Current Vessels. Retrieved from www.siferry.com/currentvessels.html, 10 September 2021.

Stauffer, N. (2020). China's transition to electric vehicles. MIT Energy Initiative, November 25. Retrieved from China's transition to electric vehicles | MIT Energy Initiative, 23 August 2021.

Strassenbahn-Online (2020). *Technische Daten der Tatra-Straßenbahnen der 2. Generation.* Retrieved from Technische Daten der Tatra-Straßenbahnen der 2. Generation (strassenbahn-online.de), 19 July 2021.

Tang, Y. (2014). Protecting our heritage. *Beijing Review.* No. 35. August 28.

US Census Bureau (n.d.a). *QuickFacts: New York city, New York; New York.* Retrieved from U.S. Census Bureau QuickFacts: New York, 31 March 2021.

US Census Bureau (n.d.b). Sex of Workers by Means of Transportation to Work for Workplace Geography. *2017 American Community Survey 1-Year Estimates.* Retrieved from American FactFinder - Results (archive.ph), 30 August 2021.

US Census Bureau (2021). *United States Commuting at a Glance: American Community Survey 1-Year Estimates.* Washington: US Census Bureau. Retrieved from www.census.gov/topics/employment/commuting/guidance/commuting/acs-1yr.html, 20 September 2021.

US Department of Energy (2020). Average fuel economy by major vehicle category. Retrieved from Alternative Fuels Data Center: Maps and Data - Average Fuel Economy by Major Vehicle Category (energy.gov), 2 September 2021.

US Department of Energy (n.d.). Emissions from hybrid and plug-in electric vehicles: Energy efficiency and renewable energy. *Alternative Fuels Data Center.* Retrieved from Alternative

Fuels Data Center: Emissions from Hybrid and Plug-In Electric Vehicles (energy.gov), 24 June 2021.

University of Exeter (n.d.). Calculation of CO_2 emissions. Retrieved from Calculation of CO_2 emissions (exeter.ac.uk), 2 September 2021.

Urban Electric Transit (2021). Vehicle Statistics: North Korea, Trolleybus. Retrieved from North Korea, Trolleybus — Vehicle Statistics — Urban Electric Transit (transphoto.org), 19 July 2021.

ViaHero (2020). The best and worst methods of transportation in Cuba. *ViaHero*. Retrieved from The Best and Worst Methods of Transportation in Cuba - ViaHero, 28 June 2021.

Wang, D. and Chai, Y. (2009). The jobs–housing relationship and commuting in Beijing, China: The legacy of Danwei. *Journal of Transport Geography 17*(1), 30–38.

Wang, J., Huang, J. and Du, F. (2020). Estimating spatial patterns of commute mode preference in Beijing. *Regional Studies, Regional Science* 7(1), 382–386.

Wang, Y., Yoshitsugu Hayashi, Y., Jin Chen, J. and Li, Q. (2014). Changing urban form and transport CO_2 emissions: An empirical analysis of Beijing, China. *Sustainability 6*, 4558–4579.

Warren, J. and Enoch, M. (2006). Mobility, energy, and emissions in Cuba and Florida. *Transportation Research Part D: Transport and Environment 11*(1), 33–44.

Warren, J., Morris, E., Enoch, M., Padilla Magdaleno, I., Parra Arias, Z. and Guanche, J. (2015). Developing an equitable and sustainable mobility strategy for Havana, *Cities 45*, 133–141.

Weiss, D.J., Nelson, A., Gibson, H.S., Temperley, W., Peedell, S., Lieber, A., Hancher, M., Poyart, E., Belchior, S., Fullman, N., Mappin, B., Dalrymple, U., Rozier, J., Lucas, T.C.D., Howes, R.E., Tusting, L.S., Kang, S.Y., Cameron, E., Bisanzio, D., Battle, K.E., Bhatt, S. and Gething, P.W. (2018). A global map of travel time to cities to assess inequalities in accessibility in 2015. *Nature 553*(7688), 333–336.

Whiting, N. (2019). Mystery solved: Why cubans call buses "Guagua". Retrieved from https://translatingcuba.com/mystery-solved-why-cubans-call-buses-guagua/, 28 June 2021.

Wikipedia (2020). Havana Suburban Railway. Retrieved from Havana Suburban Railway - Wikipedia, 24 June 2021.

Wołek, M., Wolański, M., Bartłomiejczyk, M., Wyszomirski, O., Grzelec, K. and Hebel, K. (2021). Ensuring sustainable development of urban public transport: A case study of the trolleybus system in Gdynia and Sopot (Poland). *Journal of Cleaner Production 279*, 1–14.

WorldAtlas (2021). The Boroughs Of New York City—NYC Boroughs Map. Retrieved from www.worldatlas.com/articles/the-boroughs-of-new-york-city.html, 24 September 2021.

World Population Review (2021). Population of Cities in Spain (2021). Retrieved from Population of Cities in Spain (2021) (worldpopulationreview.com), 13 July 2021.

Xiao, Z., Liu, T., Chai, Y. and Zhang, M. (2020). Corporate-run society: The practice of the danwei system in Beijing during the planned economy period. *Sustainability, 12*(4), 15.

Xinhua (2017). Beijing's electric buses to number 10,000 by 2020. *China Daily*. 2 August. Retrieved from Beijing's electric buses to number 10,000 by 2020 - China - Chinadaily.com.cn, 26 August 2021.

Yang, J., Chen, S., Qin, P. and Lu, F. (2015). The Effects of subway expansion on traffic conditions: Evidence from Beijing. *Environment for Development Discussion Paper Series*. August.

Yu, L., Shi, L. and Zhao, L. (2019). E-bikes regulation in Beijing. *Tsinghua China Law Review 11*(2), 341–368.

Yuan, X., Liu, X. and Zuo, J. (2015). The development of new energy vehicles for a sustainable future: A review. *Renewable and Sustainable Energy Review 42*, 298–305.

Zhang, J., Zhang, Y., Yang, Z., Fath, B. and Li, S. (2013). Estimation of energy-related carbon emissions in Beijing and factor decomposition analysis. *Ecological Modelling 252*, 258–265.

Zheng, X. (2018). *China's 40 Years of Economic Reform and Development*. Singapore: Springer.

Zhou, J., Wang, M. and Long, Y. (2017). Big data for intrametropolitan human movement studies. *International Review for Spatial Planning and Sustainable Development 5*(3), 100–115.

6

CONSUMER GOODS

6.1 Introduction

The economic effect of cities has already been noted in Chapter 1. Living in cities in turn tends to lead to a larger personal EF than the national average. This chapter investigates and compares the consumption of consumer goods in the 19 cities. The first problem is defining what consumer goods are. The Organisation for Economic Co-operation and Development (OECD) survey of consumer goods and services collects "the prices that purchasers actually pay to sellers to acquire the goods and services specified" (OECD/Eurostat, 2012:109). The specified services fall into the six groups of Food (including drinks and tobacco); Personal Appearance (clothing, shoes, care of same, care of appearance); House and Garden (including hobbies and toys); Transport, Restaurants and Hotels (including camping and sports equipment); Services (including car and house maintenance, veterinary care, and telecommunication services); and Furniture and Health (including medical products). However, the OECD study is concerned with total household expenditure, so the definition used for consumer goods is very wide. The typical dictionary definition of consumer goods is goods bought to satisfy consumer needs, rather than goods bought by manufacturers for the purpose of making something, but this is not helpful in defining the difference between essential household expenditure, such as on food, and what might be considered expenditure on consumer goods and services, such as buying a television or a ticket to the cinema.

Clothing illustrates the divide between what is and what is not essential. Most societies require people to be clothed, but at the same time having more clothes than are strictly needed puts clothing in the consumer goods category. The modern global fashion market accounts for around US$3 trillion annually or about 2% of global GDP. "It is estimated that around the world, about 107 billion units of apparel and 14.5 billion pairs of shoes were purchased in 2016. That equates to every person

DOI: 10.4324/9781003045847-7

TABLE 6.1 Annual money spent on apparel for selected nations US$/person (CHIC, 2019)

Nation	*City*	*Annual spend (US$)*	*No. items bought*	*Average cost per item (US$)*
USA	New York	844	53	15.9
China	Beijing	190	30	6.3
Brazil	São Paulo	117	11	10.6
India	Ahmedabad	34	5	6.8
Egypt	Cairo	27	2	13.5

on the planet buying roughly 13 garments and two pairs of shoes" (Fashion United, 2021). Using data from Euromonitor International, Table 6.1 compares this global statistic with what was spent for selected cities, although the date of this spend is not known, showing that as might be expected more clothing is purchased in a wealthy city like New York than in Cairo, even accounting for differences in climate.

Consumer goods also tend to be thought of as objects that have been made for sale to someone else. Wackernagel and Rees (1996:82), in their breakdown of the components of an EF, defined consumer goods as "packaging, clothing, furniture and appliances, books and magazines, tobacco and alcohol, personal care, recreation equipment and other goods." This broadly represents what each person spends once they have accounted for the recurring costs of their food, shelter and transport, although it does not account for money spent on tourism or money spent on private education and health. As finding comparative data for the 19 cities is not easy, aspects of the Wackernagel and Rees definition have been examined below as well as a comparison of household spending power. Most data are also national rather than city based.

6.2 Household spending power

As a means of understanding how much cash might be available to spend on consumer goods as identified by Wackernagel and Rees, Table 6.2 sets out the cities and their nations along with the type of economy as defined by the World Bank (n.d.) for a 2021 baseline. A high-income economy had a GNI per capita of at least $12,535, an upper-middle income one of $4,046–12,535, a lower-middle income one of $1,036–4,045 and $1,035 and below is classed as a low-income economy. These values have been applied to the cities in Table 6.2 together with the World Bank (2021a) assessment of GNI and the date to which this applies (note these values differ slightly from the United Nations values in Table 1.8). Because countries tax their citizens at different rates, a further column has been added for disposable income in US dollars where this is known, taken from the OECD Better Life Index (OECD, n.d.). The relationship between disposable income, which is what a household has to spend, including on buying consumer goods, and tax rates is important. Higher tax rates mean more services are supplied by governments,

TABLE 6.2 Economy type based on income

City	*Nation*	*Income type*	*GNI*	*Disposable income (US$)*
New York	USA	High	65,850 (2019)	45,284
Copenhagen	Denmark	High	64,000 (2019)	29,606
Singapore	Singapore	High	59,590 (2019)	
Sydney	Australia	High	55,100 (2019)	32,759
Amsterdam	Netherlands	High	53,060 (2019)	29,333
Hamburg	Germany	High	48,600 (2019)	34,294
Dubai	UAE	High	43,470 (2019)	
Auckland	New Zealand	High	42,220 (2019)	
Moscow	Russia	Upper-middle	11,260 (2019)	
Kuala Lumpur	Malaysia	Upper-middle	11,230 (2019)	
Buenos Aires	Argentina	Upper-middle	11,130 (2019)	
Beijing	China	Upper-middle	10,390 (2019)	
São Paulo	Brazil	Upper-middle	9,130 (2019)	
Havana	Cuba	Upper-middle	7,480 (2016)	
Jakarta	Indonesia	Upper-middle	4,050 (2019)	
Cairo	Egypt	Lower-middle	2,690 (2019)	
Ahmedabad	India	Lower-middle	2,120 (2019)	
Nairobi	Kenya	Lower-middle	1,750 (2019)	
Pyongyang	North Korea	Low	1,241 (2019)*	

* (Yoon, 2021).

especially in areas like health and education, which will be reflected in what a household has to purchase. This becomes apparent in the discussion of the individual cities.

However, Table 6.2 does little to explain how people spend their income. As noted above, this also depends on what the state supplies in terms of health care and education or whether these are charges on personal income.

Another set of income categories, developed by Kearney (n.d.) and based much earlier in 2010, gives a sense of which items people buy as their incomes increase. In this analysis, some 25% of the world population fell into the "basic" category, where most disposable income is spent on food and very little on leisure and entertainment. The "emerging" category covered more than 25% of the global population and although food again dominated spending, there was more money for clothing and footwear and transportation. The "escalating" category applied to 600 million people and their spending was much more balanced between food, transport, personal property and leisure and entertainment. Twice as many people (1.2 billion) fell into the "established" consumer group, where a significantly smaller proportion of disposal income is spent on food. Kearney (n.d.) notes, "Established Consumers spend significant amounts of money on personal care, leisure, entertainment, and healthcare (in countries where the government does not provide it)." The problem is that many countries are moving upwards through these categories as disposable incomes increase, with consequent pressure on global resources to satisfy the resulting consumer cravings.

To investigate consumer spending in more detail, the following sections look at a selection of cities for which data could be found, based on the ranking in Table 6.2, noting that most statistics are nationally rather than city based. Comparisons are then made in the final section.

6.3 New York

The US Bureau of Labor Statistics (2021) gives the 2018/19 average annual New York income before taxes as $87,714, ranging from $11,718 for the lowest 20% of citizens to $226,926 for the highest 20%. The average annual expenditure for the same dates was $65,712, ranging from $26,440 for the lowest 20% to $131,529 for the highest 20%. Table 6.3 sets out how the average annual disposable income was distributed as both percentage and monetary values. The values in italics are those that would be considered consumer goods. Communications have been placed in this category as the phone is normally found in tables of consumer goods ownership. Small differences in total values are due to rounding. For those with very low incomes, subsidies are available, such as the housing subsidy (New York State, n.d.), which is why expenditure is greater than income in some of the figures above.

In Table 6.3, 20% of the total could be considered expenditure on consumer goods. To probe further into what households own, Table 6.4 looks at household ownership of some large appliances (StatInvestor, 2021) and IT devices (United States Census Bureau, 2021), with the date of the available data. As expected in a wealthy society, ownership of appliances and goods is high.

TABLE 6.3 Distribution of 2018/19 average disposable income, New York

Category	*% disposable income*	*US$/person*
Food*	14.0	9,269
Alcohol and tobacco	1.4	947
Housing (including water and maintenance)	24.2	16,004
Energy	3.7	2,416
Communications (internet, telephone, postage)	2.3	1,525
Household furnishi ngs and equipment	2.9	1,887
Household operations and housekeeping supplies	3.5	2,328
Clothing and footwear	3.8	2,483
Transport	13.1	8,616
Health care	6.6	4,338
Entertainment (pets, hobbies reading)	4.9	3,221
Personal care	1.2	817
Education	3.2	2,122
Insurance services and pensions	11.7	7,733
Other	3.5	2,298

* 57% consumed at home and 43% away from home.

TABLE 6.4 Household appliance and device ownership, USA

Appliance/goods	*% household ownership*
Smartphone	84 (2018)
Desktop/laptop computer	78 (2018)
Tablet	63 (2018)
Dishwasher	73 (2017)
Washing machine	86 (2017)
Dryer	84 (2017)
Refrigerator	98 (2017)
Microwave	92 (2017)

6.4 Copenhagen

Copenhagen is in the Hovedstaden region of Denmark, one of five regions that administratively sit between central and municipal government. When it comes to consumption, the 2019 total for Hovedstaden, the highest of the five provinces, was an average 354,547 DKK/household (Statistics Denmark, n.d.a) (US$58,110). However, the income tax rate is high in Denmark at an average 45% of income as the state uses this money to ensure a good quality of life for all Danes. "In Denmark, it is widely considered that the vast benefits afforded by the welfare system greatly outweigh the high Danish taxes" (IRIS, 2021).

Using national averages, the 2019 Danish household expenditure is broken down as shown in Table 6.5 (Statistics Denmark, n.d.b). The household cost (column three) is based on the average household income for Hovedstaden less

TABLE 6.5 Breakdown of 2019 average expenditure, Copenhagen

Category	*% household expenditure*	*US$/household*	*US$/person*
Food	11.7	4,419	2,104
Alcohol/tobacco	2.1	793	378
Clothing/footwear	3.7	1,398	666
Housing (including water and maintenance)	26.0	9,821	4,677
Energy	7.3	2,757	1,313
Household (including furniture, appliance, tools, maintenance)	4.9	1,850	881
Medical products and services	2.4	907	432
Transport (including buying and operating cars)	13.9	5,250	2,500
Communications	2.5	944	450
Recreation (including pets, newspapers, holidays, hobbies)	10.0	3,777	1,799
Education (including pre-school)	0.5	189	90
Catering and accommodation services	6.1	2,304	1,097
Personal care/insurance/other personal effects	8.8	3,324	1,583

TABLE 6.6 Danish appliance and consumer goods ownership

Appliance/goods	*% household ownership 2020*	*% household ownership 2021*
Smartphone	90	90
Tablet	61	59
Portable computer	88	88
Dishwasher	75	77
Washing machine	81	83
Dryer	56	58

income tax at 45% (US$37,772). The 2019 national average household size of 2.1 persons (Esri, 2021) has been applied for an approximate cost per person in column four.

Because of the welfare system, little disposable income is given to education or health, with housing accounting for the largest outlay, followed by transport and food. When it comes to consumer goods (shown in italics), these account for 19.3% of outlay.

Looked at another way, Table 6.6 sets out the 2020 and 2021 percentage of households with various named consumer goods (Statistics Denmark, n.d.c). Like New York, ownership levels are high in a wealthy city like Copenhagen.

6.5 Singapore

Data in this section have been taken from the 2017/2018 household expenditure survey by the Singapore Department of Statistics (2019). For the stated period, food, housing and transport accounted for 62% of monthly expenditure, and "Households spent an average of $4,906 [US$3700.5] a month on goods and services" (Singapore Department of Statistics, 2019:1). This gives an annual figure of US$44,406. This value has been used to give a breakdown of annual cost per household in column three of Table 6.7. The same source gives the 2017/18 monthly expenditure per household member as $1,628 (US$1228), or an annual figure of US$14,736. This has been applied in column four to give an annual cost per person.

From Table 6.7, 13.8% of disposable income goes towards what might be termed consumer goods in Singapore, considerably less than the approximate 20% of New York and Copenhagen. Singaporeans spend more on food (20.3%) than New York (14%) and Copenhagen (11.7%).

TABLE 6.7 Breakdown of household expenditure for 2017/8 (Singapore)

Category	*% household expenditure*	*US$/household*	*US$/person*
Food	20.3	9,014	2,991
Clothing/footwear	2.1	933	309
Housing and related expenditure	28.9	12,8333	4,259
Health	5.5	2,442	810
Transport	13.2	5,862	1,945
Communications	4.1	1,821	604
Recreation and culture	6.4	2,842	943
Educational services	5.7	2,531	840
Accommodation services	1.2	533	177
Other	12.6	5,595	1,857

TABLE 6.8 Percentage household ownership of consumer goods (Singapore Department of Statistics, 2019:55)

Appliance/goods	*% household ownership (2017/18)*
Air conditioner	79.7
Mobile phone	98.0
Desktop computer*	31.7
Tablet	47.8
Laptop	69.1
Television	97.0
Dryer	19.2
Washing machine	96.1
Vacuum cleaner	74.4
Massage chair	8.3

* 81.4% of households had some type of computer (Singapore Department of Statistics, 2019:69).

6.6 Pyongyang

Official data on Pyongyang is unavailable, but from the low GDP in Table 1.8, it is apparent that consumer goods are not going to feature large in household expenditure. Some idea of what consumer goods might be found comes from French (2014).

> According to defectors, North Koreans want "five chests and seven appliances". The chests are a quilt chest, wardrobe, bookshelf, cupboard and shoe closet, while the appliances comprise a TV, refrigerator, washing machine, electric fan, sewing machine, tape recorder and camera. Most ordinary people only have a couple of appliances, usually a television and a sewing machine.

The same source also states electricity supply is not continuous and one side of a street may be lit until switching-over time when the electricity supply is passed to the other side. This would make owning a refrigerator problematic, however desirable. "Housing is free in North Korea but is often cramped and lacking amenities. Extended families average 4.4 persons while most households live in two-room units" (Haub, 2010), although this statement is based on the 2008 census, then only the second to be carried out. The US Library of Congress (n.d.) states the average family consists of two generations and has four to five persons. A more recent visitor to North Korea stated that in "model homes, residents had all furniture and furnishings assigned by the state, and kept their bath and spare buckets filled with water because supply is intermittent" (Graham-Harrison, 2018b). Goods are also often in short supply. Holidays are rare and the several cinemas in Pyongyang are often shut because of the lack of power. TVs and radios have to be registered to ensure they are tuned to the allowed stations (French 2014). In the DPRK as a whole in 2019, 4.6% of the population subscribed to a landline and 15% to a mobile cellular network, with the launch of the mobile telephone service in 2008 in Pyongyang (CIA, 2021), suggesting use of mobile phones might be higher in the capital than in the country as a whole. For recreation, in or near Pyongyang, there are monuments, a science museum, parks and a water park for use in summer (Graham-Harrison, 2018a). Sunday is the day for walking as, just as in many countries in the developed world, public transport services are reduced (French, 2014), though the roads are never crowded given private car ownership is allowed but "highly restricted" (Kang, 2017). North Korea is producing versions of Western fashion brands such as Nike and Adidas, but these are only available to the elite in society, who are also the people who can afford to use a taxi. "The fare for a single ride starts at around 16,000 won, equivalent to just under $2 at the black-market exchange rate used by North Koreans, but around a month's wages for many workers in the capital" (Graham-Harrison, 2018b). The same source goes on to state that about 10% of the population could be described as the new middle class and only a small fraction of this would be the elite who could afford a taxi or a pair of North Korean made Adidas trainers.

This suggests that expenditure on consumer goods services is going to form a small proportion of disposable income for all except the elite in Pyongyang. To get a sense of costs, Table 6.9 sets out the cost of items relative to their cost in the United States in columns one and two, based on advice for expats living in North Korea (Xpatulator, n.d.), showing that with the exception of clothing and communication, consumer goods in the DPRK are more expensive than in the United States. As a guide for real costs, columns three and four list specific items that fall within some of the groupings and their costs for tourists (Lonely Planet, 2021).

In 2019 the average income in DPRK was said to be "less than 4%—or one-twenty-seventh—of what the average South Korean makes" (Kim, 2020). In 2019 the annual average wage in South Korea was US$42,285 (Statista, n.d.), making the equivalent in the DPRK US$1,566. Income tax was abolished in 1974, but indirect taxes are still collected by the government: "Fees are charged to farmers for seeds, fertiliser, irrigation water, and equipment. Consumers pay a tax for the use of water and certain other household amenities" (Nations Encyclopedia, 2021).

The environmental impact of the consumer goods of the average citizen of Pyongyang is not likely to be high, and virtually zero in the rest of the

TABLE 6.9 % difference in cost between DPRK and the United States and cost of some items

Item	*% difference with the United States*	*Item*	*Cost (US$)*
Alcohol/tobacco	+26	Local beer	60 cents
Clothing	–30		
Communication	–5	KoryoLink SIM card	257
Education	+6		
Furniture and appliances	+16		
Groceries (including cleaning products)	+6	Street snack	12 cents
Healthcare	–30		
Accommodation	–71		
Personal care	+1		
Recreation and culture (cinema, books, sports)	+43	Cheapest mass games ticket	95
Meals out/*hotels*/take away	–34	Nightly rate for medium hotel	120–180
Transport	+20	Metro ticket	1 cent
Miscellaneous (dry cleaning, postage, newspapers etc.)	+101		

TABLE 6.10 Estimated annual expenditure per person (Pyongyang)

Item	*%*	*US$/household*	*US$/person*
Alcohol/tobacco	1	27	6
Clothing	3	81	18
Communication	1	27	6
Education	-	-	-
Furniture and appliances	1	27	6
Food	90	2,466	548
Health care	-	-	-
Accommodation	-	-	-
Personal care	1	27	6
Recreation and culture (cinema, books, sports)	1	27	6
Meals out/*hotels*/take away	-	-	-
Transport	1	27	6
Miscellaneous (dry cleaning, postage, newspapers etc.)	1	27	6

country: "Consumerism is almost non-existent away from the capital. With very little disposable income, most North Koreans spend any extra cash they earn on food for their families" (Love Money, n.d.). This position will have intensified with the recent shortage of food in the DPRK (Yoon, 2021). Table 6.10 is thus a possibly fictitious estimate of how the average household disposable income of US$2,740 might be spent, based on 1.75 working people in a 4.5-person family (figures rounded). The proportion of women in the DPRK labour force was an average of

75% for the period 1999–2019 (Global Economy, 2021). The table also draws on the discussion above and the analysis of the other three low-income cities (Nairobi, Ahmedabad and Cairo) that follows.

With only 9% of income going into consumer goods, including clothing, which in a cold climate might be more of a necessity than a luxury, this makes Pyongyang the city with the lowest expenditure on this EF category.

6.7 Nairobi

A 2009 survey of households in Nairobi found that in comparison with a 2003 survey, an increasing proportion of disposable income was being spent on food (Kamau et al., 2011). Table 6.11 sets out the income quintiles, monthly household expenditure, annual expenditure and the percentage spent on food in 2003 and 2009.

On average, 29% of 2009 household expenditure was on food, though this is skewed by the low percentage on the highest income quintile. Everyone is spending more on food except the very rich who are spending less. The four lower quintiles also spend a far greater percentage of their income on food, compared to the wealthiest. A much earlier 1982–83 survey of low-income households in Nairobi found the following breakdown of consumer expenditure (Central Bureau of Statistics: Kenya Statistical Digest (Nairobi), n.d.) (Table 6.12), with a much higher proportion of income spent on food compared with 2003 but quite similar to the situation in 2009.

The analysis of a 2015–16 household survey for the whole of Kenya found "Richer households tend to be smaller (4.4 household members in the top expenditure quintile as compared to 6.8 in the lowest), located in urban settings and report considerably higher availability levels of modern amenities" (Ilinca et al., 2019). In the absence of any more up to date data, an average household size of 5.6 along with the breakdown in Table 6.12 has been applied to the 2009 average annual expenditure from Table 6.11 to give a breakdown of spending per person in Table 6.13.

Omitting education, 12.4% of income is spent on consumer goods. A study based on a 2009 dataset found 67.5% of urban dwellers in the sample had access

TABLE 6.11 2009 household expenditure (Nairobi)

Quintile	*Monthly expenditure (US$)*	*Annual expenditure (US$)*	*% spent on food 2003*	*% spent on food 2009*
Lowest	130	1,560	25	49
Second	177	2,124	33	44
Third	234	2,808	36	41
Fourth	378	4,536	30	34
Highest	1,306	15,672	25	16
Average	445	5,340	30	37

TABLE 6.12 Breakdown of 1982–83 low-income household expenditure

Category	*% expenditure*
Food	44.2
Beverages and tobacco	2.1
Clothing and shoes	5.0
Rent	25.0
Fuel, light, water	3.1
Furniture, furnishings, household equipment and operation	5.3
Health and personal care	3.0
Transport and communication	4.1
Education, *recreation and entertainment*	6.2
Miscellaneous goods and services	2.0

TABLE 6.13 Annual expenditure per person (Nairobi)

Category	*% expenditure*	*Annual expenditure per household (US$)*	*Annual expenditure per person (US$)*
Food	44.2	2,360	421
Beverages and tobacco	2.1	112	20
Clothing and shoes	5.0	267	48
Rent	25.0	1,335	238
Fuel, light, water	3.1	166	30
Furniture, furnishings, household equipment and operation	5.3	283	51
Health and personal care	3.0	160	29
Transport and communication	4.1	219	39
Education, *recreation and entertainment*	6.2	331	59
Miscellaneous goods and services	2.0	107	19

TABLE 6.14 Estimated appliance ownership (Nairobi)

Appliance	*% Nairobi*
Iron	47
Refrigerator	24
Television/radio	61

to electricity (Mbaka et al., 2019). Penetration in 2019 for the whole of Kenya was only marginally higher at 69.7% (The World Bank, 2021b). Magamba (2010:7) gives a breakdown of appliance use for households with electricity as 99% for lighting, 90% for entertainment (TV and radio), 69% ironing and 35% refrigeration. Combining these two sets of figures for electricity supply and usage, gives the ownership values in Table 6.14.

6.8 Ahmedabad

In 2017–18, the per capita income of Gujarat, in north-west India where Ahmedabad is located, was INR 174,652 (US$2,381) (Jagran Prakashan Ltd, 2021), whereas the average per capita gross national income for that period was INR 128,497 (US$1,752) (MOSPI, n.d.). Another source (Statista, 2021) gives the estimated 2018 gross national income as INR 114,960 (US$1,567). The same source goes on to note that "a small wealthy class of people in the country can skew the per capita income substantially, even though the average population has no change in income. This is exemplified by the fact that in India, the top one percent of people control over 58 percent of the country's wealth." This suggests that trying to find out how much citizens of Ahmedabad, Gujarat, or even India, spend on consumer goods is going to be complicated.

A report on the future of Indian consumers (World Economic Forum, 2018:13) divides India into five consumer types with suggested annual household incomes (Table 6.15).

It should be no surprise that the rural poor in India are unlikely to own the major consumer goods, not just because their money is spent on more basic items such as food, education and clothing, but because the supply of electricity is not necessarily available or reliable (Phadke et al., 2019), and without a secure supply of electricity, a refrigerator is useless. However, this is not the case in a city like Ahmedabad. Ahmedabad is viewed as a relatively wealthy city and is stated to be "a successful example of inclusive urbanisation, seen in a decline in urban poverty rate from 28% to 10% and slum settlement populations from 25.6% to 4.5% in less than 20 years" (Rathore, 2015). Table 6.16 sets out the percentage breakdown of monthly household spending in Ahmedabad (Numbeo, 2021) as an estimate for those moving to Ahmedabad, rather than local statistics. However, as this is the only source available it has been used here. The same source gives the average monthly costs for a family of four living in Ahmedabad as INR 87,693.1 (US$1,182.7), which would be US$14,192/annum. This has been applied to Table 6.16 to give the annual cost for the four-person household and the annual cost per person (numbers rounded). This breakdown does not include any spending on education, which is

TABLE 6.15 Indian consumer types

Consumer type	*Average annual household income (US$)*	*% population 2018*	*% Access to smartphones*
Poor rural	2,800	29	5
Poor dreamers	5,200	14	10
Young and savvy	5,200	11	65
Middle India	6,800	19	10
Connected aspirants	6,800	15	95
Conservative rich	40,000	4	15
Sophisticated rich	40,000	7	95

TABLE 6.16 Breakdown of household expenditure (Ahmedabad)

Spending category	*% household budget*	*Annual household spend (US$)*	*Per person spend (US$)*
Markets (food, alcohol and tobacco)	43.9	6,230	1,558
Restaurants (including alcohol)	12.1	1,717	429
Clothing and shoes	3.5	497	124
Rent	16.7	2,370	593
Utilities	4.5	639	160
Transportation	12.2	1,731	433
Sport and leisure	7.1	1,007	252

TABLE 6.17 Ownership of key consumer goods (Ahmedabad)

Appliance or good	*% ownership India*	*% ownership Ahmedabad*
Washing machine	8.8	9.0
Refrigerator	21.0	49.0
Television	60.0	88.0
Air conditioner	3.0	30.0 (2018)*

* Tathagat et al. (2020).

free but only for children aged 6–14 (Angloinfo, 2021). Additionally, alcohol availability is a state issue in India and Gujarat has banned alcohol (Anantha Eashwar et al., 2020), although drinking "is deep rooted among several communities in the state. It is a custom in these communities, where both men and women sit together and drink on special occasions" (Kaushik, 2021).

Assuming the household in Table 6.16 does not drink alcohol, only 10.6% of spending goes towards consumer items, slightly higher than the 9% of Pyongyang, with the majority of spending is on food of some type.

Table 6.17 sets out the penetration of key consumer durables in the 2016 Indian market as a percentage of the whole population (Sarangi, n.d.) compared with figures for three key items in Ahmedabad (Anon, n.d.).

As expected, apart from washing machines ownership, in the wealthier city with its supply of electricity, appliance ownership is higher than the average and the hot, dry climate would explain the high ownership of air-conditioners.

6.9 Cairo

In the absence of statistics relating to Cairo, Table 6.18 sets out the 2015 percentage breakdown of annual average household expenditure for Egypt for both rural and urban households, the latest available data (CAPMAS,2018:136). According to CAPMAS, the average household income in 2015 was 44,193.8 EGP (US$2,818) (CEIC data, 2021). In 2014 the average number of people per household was 4.1

TABLE 6.18 Breakdown of 2015 household expenditure (Cairo)

Category	*Rural (%)*	*Urban (%)*	*US$ per household*	*US$ per person*
Food and non-alcoholic beverages	38.9	30.4	857	209
Alcohol/tobacco/narcotics	4.9	4.5	127	31
Clothing and shoes	6.1	5.3	149	36
Housing, water, electricity, gas and other fuels	15.8	19.0	535	131
Furnishings, household equipment and household maintenance	4.2	4.0	113	27
Health	10.0	10.1	285	69
Transport	5.1	7.4	209	51
Communication	2.0	2.9	82	20
Recreation and culture	1.9	2.2	62	15
Education	3.3	6.1	172	42
Restaurants and hotels	3.2	4.7	132	32
Miscellaneous goods and services	3.7	4.0	113	27

TABLE 6.19 2015 % ownership of selected goods and appliances

Item	*Rural (%)*	*Urban (%)*
Refrigerator	93.5	96.9
Freezer	11.4	20.2
Washing machine	10.0	45.7
Vacuum cleaner	12.0	36.1
Air conditioner	4.1	19.4
Colour TV	92.7	96.2
PC/laptop	20.9	45.5
Mobile phone	89.6	86.4

(UN, 2017:14). These values have been applied to the urban breakdown to give the US dollar value per household in column three, and per person in column four. This shows 24% of disposable income could be attributed to consumer goods if restaurants and hotels are included.

Table 6.19 sets out the 2015 ownership of appliances for both rural and urban Egypt (CAPMAS, 2018:143–144).

6.10 EF of consumer goods for the seven cities

Table 6.20 sets out the EFs of the seven countries (GFN, 2021) and the annual money spent per person on consumer goods (categories in italics in the previous tables). Assuming that GNI per person represents a country's final income before tax, then this can stand for the whole of the EF, and the after-tax spending on

TABLE 6.20 Comparison of consumer goods and resulting ecological footprint for seven cities

City	*National EF (2017)*	*US$/annum on consumer goods*	*GNI (2019)*	*EF consumer goods*
New York	8.1	13,208	65,850	1.6
Copenhagen	6.9	6,368	64,000	0.7
Singapore	5.9	1,930	59,590	0.2
Pyongyang	0.8	54	1,241	0.03
Nairobi	1.0	178	1,750	0.1
Ahmedabad	1.2	376	2,120	0.2
Cairo	1.8	161	2,690	0.1

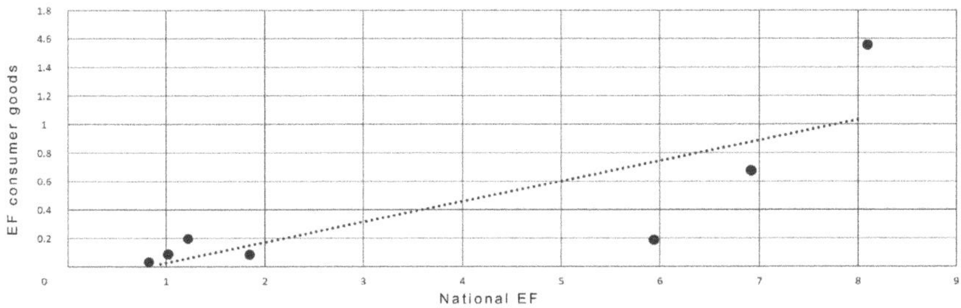

FIGURE 6.1 Relationship between national EF and EF of consumer goods

consumer goods represents that portion of the EF related to these goods. Taxes claimed by governments will be used for education and health provisions among others, supplemented by additional spending from disposable income as seen in the previous tables. However, here the only interest is in the EF of consumer goods. Column four sets out the GNI per person from Table 6.2, and column five shows the EF of consumer goods based on the ratio of annual consumer goods spending to GNI. Worthy of comment is the fact the EF of consumer goods in New York is the same as the sustainable EF for all people of the world (see p.183).

Plotting these results gives the following scatter plot and trend line (Figure 6.1). As might be expected the wealthier the nation, the higher the EF and the higher the EF of consumer goods.

References

Anantha Eashwar, V.M., Umadevi, R. and Gopalakrishnan, S. (2020). Alcohol consumption in India: An epidemiological review. *Journal of Family Medicine and Primary Care 9*(1). 49–55.

Angloinfo (2021). The School System in India. Retrieved from www.angloinfo.com/how-to/india/family/schooling-education/school-system, 21 June 2021.

Anon (n.d.). India's Consumer Durables Market. Retrieved from https://images.livemint.com/r/LiveMint/Period2/2018/08/09/Photos/Processed/plainfacts1.jpg, 16 June 2021.

CAPMAS (Central Agency for Public Mobilisation and Statistics) (2018). Egypt in Figures. Retrieved from www.sis.gov.eg/UP/Egypt%20in%20Figures%202018/egypt-in-numbers2018.pdf, 23 June 2021.

CEIC data (Census and Economic Information Centre) (2021). Egypt Average Household Income: Value. Retrieved from www.ceicdata.com/en/egypt/average-household-income/average-household-income-value, 23 June 2021.

Central Bureau of Statistics: Kenya Statistical Digest (Nairobi) (n.d.). Consumer Price Index—Lower Income Group. Retrieved from http://laborsta.ilo.org/applv8/data/SSM1/E/KE.html, 21 June 2021.

CIA (Central Intelligence Agency) (2021). *The World Factbook—Korea, North*. Retrieved from www.cia.gov/the-world-factbook/countries/korea-north/, 23 June 2021.

CHIC (China International Fashion Fair) (2019). The Size of the *Global Fashion Retail Market*, USA and China Spend by far the Most on Apparel. Retrieved from https://enmar.chicfair.com/syzhxw/8275.html, 3 June 2021

Esri (Environmental Systems Research Institute) (2021). Average Household Size in Denmark. Retrieved from www.arcgis.com/home/item.html?id=2fe5f2747b9c452dba17f52d78d54ec, 14 June 2021.

Fashion United (2021). Volume and Consumption: How Much Does the World Buy? Retrieved from www.commonobjective.co/article/volume-and-consumption-how-much-does-the-world-buy, 3 June 2021.

French, P. (2014). A day in the life of Pyongyang—how North Korea's capital goes to work. Retrieved from *The Guardian*, 22 June 2021.

The Global Economy (2021). North Korea: Female labor force participation. Retrieved from www.theglobaleconomy.com/North-Korea/Female_labor_force_participation/, 24 June 2021.

GFN (Global Footprint Network) (2021). World Map. Retrieved from https://data.footprintnetwork.org/#/, 24 June 2021.

Graham-Harrison, E. (2018a). A glimpse of daily life in Pyongyang—in pictures. *The Guardian*, 22 June 2021.

Graham-Harrison, E. (2018b). Fake Adidas and Hello Kitty: a taste of consumerism for North Korea's elite. Retrieved from *The Guardian*, 22 June 2021.

Haub, C. (2010). North Korea Census Reveals Poor Demographic and Health Conditions. Retrieved from www.prb.org/resources/north-korea-census-reveals-poor-demographic-and-health-conditions/#:~:text=Housing%20is%20free%20in%20North,live%20in%20two%2Droom%20units, 23 June 2021.

Ilinca, S., Di Giorgio, L., Salari, P. and Chuma, J. (2019). Socio-economic inequality and inequity in use of health care services in Kenya: evidence from the fourth Kenya household health expenditure and utilization survey. *International Journal for Equity in Health, 18*, 1–13.

IRIS (2021). Why Danes Appreciate the High Danish Income Tax Rate. Retrieved from https://fmpglobal.com/blog/why-danes-appreciate-the-high-danish-income-tax-rate/#:~:text=The%20Danish%20Income%20Tax%2 0Rate,28.9%25%20of%20GDP%20in%202019, 10 June 2021.

Jagran Parkashan Ltd. (2021). Per Capita Income of Indian States 2019–20. Retrieved from Per Capita Income of Indian States 2019-20 (jagranjosh.com), 15 June 2021.

Kamau, M., Olwande, J. and Githuku, J. (2011). Consumption and Expenditures on Key Food Commodities in Urban Households: The Case of Nairobi. Tegemeo Institute of Agricultural Policy and Development. Retrieved from www.tegemeo.org/images/_tegemeo_institute/downloads/publications/working_papers/wp41a.pdf, 21 June 2021.

Kang, M-J. (2017). Car ownership in North Korea on the rise. Retrieved from www.dailynk.com/english/car-ownership-in-north-korea-on-th/, 22 June 2021.

Kearney (n.d.). Consumer Wealth and Spending: The $12 Trillion Opportunity. Retrieved from www.es.kearney.com/consumer-retail/article/?/a/consumer-wealth-and-spending-the-12-trillion-opportunity, 4 June 2021.

Kaushik, H. (2021). In dry Gujarat, women drinkers double in 5 years, percentage of drinking men slide by 50%. *The Times of India*. Retrieved from https://timesofindia.indiatimes.com/city/ahmedabad/in-dry-gujarat-women-drinkers-double-in-5-yrs-percentage-of-drinking-men-slide-by-50/articleshow/80205515.cms, 21 June 2021.

Kim, J. (2020). North Koreans earn just 4% of the average South Korean's income, new report says. Retrieved from www.nknews.org/2020/12/north-koreans-earn-just-4-of-the-average-south-koreans-income-new-report-says/, 22 June 2021.

Lonely Planet (2021). North Korea in Detail: Money and Costs. Retrieved from www.lonelyplanet.com/north-korea/a/nar-gr/money-and-costs/357180, 23 June 2021.

Love Money (n.d.). The shocking inequality between Kim Jong-un and his people. Retrieved from www.lovemoney.com/gallerylist/62439/the-shocking-inequality-between-kim-jongun-and-his-people, 23 June 2021.

Magambo, C.K. (2010). A Study of Household Electricity Demand and Consumption Patterns in Nairobi. Master of Science Thesis. University of Nairobi. Retrieved from http://erepository.uonbi.ac.ke/bitstream/handle/11295/4184/Magambo_A%20Study%20of%20Household%20Electricity%20Demand%20and%20Consumption%20Patterns%20in%20Nairobi.pdf?sequence=1&isAllowed=y, 21 June 2021.

Mbaka, C.K., Gikonyo, J. and Kisaka O.M. (2019). Household's energy preference and consumption intensity in Kenya. *Energy, Sustainability and Society 9,* 20.

MOSPI (Ministry of Statistics and Programme Implementation) (n.d.). Statement 1.2: Per Capita Income, Product and Final Consumption. Retrieved from http://mospi.nic.in/sites/default/files/reports_and_publication/statistical_publication/National_Accounts/NAS19/S1.2.pdf, 15 June 2021.

Nations Encyclopedia (2021). Korea, Democratic People's Republic of (DPRK)—Taxation. Retrieved from www.nationsencyclopedia.com/Asia-and-Oceania/Korea-Democratic-People-s-Republic-of-DPRK-TAXATION.html, 23 June 2021.

New York State (n.d.). Temporary Assistance. Retrieved from https://otda.ny.gov/programs/temporary-assistance/, 14 June 2021.

Numbeo (2021). Cost of Living in Ahmedabad. Retrieved from www.numbeo.com/cost-of-living/in/Ahmedabad, 21 June 2021.

OECD (n.d.). Better Life Index. Retrieved from https://stats.oecd.org/Index.aspx?DataSetCode=BLI, 10 June 2021.

OECD/Eurostat (2012). Consumer goods and services, in Eurostat-OECD *Methodological Manual on Purchasing Power Parities*. Paris: OECD Publishing.

Phadke, A., Park, Y.W. and Abhyankar, N. (2019). Providing reliable and financially sustainable electricity access in India using super-efficient appliances. *Energy Policy 132,* 1163–1175.

Rathore, V. (2015). Ahmedabad is no city for the poor, says study. Retrieved from https://thewire.in/economy/ahmadabad-is-no-city-for-the-poor-says-study, 17 June 2021.

Sarangi, P. (n.d.). The Indian Consumer Durable Market and an Analysis of Demand Pattern for Major Durables. Retrieved from Major Durables.pdf (icsi.edu), 17 June 2021.

Singapore Department of Statistics (2019). Report on the Household Expenditure Survey 2017/2018. Retrieved from www.singstat.gov.sg/-/media/files/publications/households/hes201718.pdf, 10 June 2021.

StatInvestor (2021). Device Ownership Levels in U.S. Households 2017. Retrieved from https://statinvestor.com/data/34902/household-appliancedevice-ownership-in-us-households/, 14 June 2021.

Statista (n.d.). Average annual wage in South Korea adjusted for Purchasing Power Parity from 2000 to 2019. Retrieved from www.statista.com/statistics/557759/south-korea-average-annual-wage/, 22 June 2021.

Statista (2021). Per capita national income across India in financial year 2015 and 2017, with estimates until 2021. Retrieved from India: national income per capita 2015–2021 | Statista, 16 June 2021.

Statistics Denmark (n.d.a). Household Budget Survey. Retrieved from Household Budget Survey - Statistics Denmark (dst.dk), 8 June 2021.

Statistics Denmark (n.d.b). Consumption by price unit, households, group of consumption and time. Retrieved from Statistikbanken (statbank.dk), 8 June 2021.

Tathagat, T., Varma, P., Prasad, A., Shukla, N., Walia, A. and Dhingra, N. (2020). Unraveling Urban Household Energy Use Patterns in India. Retrieved from www.clasp.ngo/wp-content/uploads/2021/04/Unraveling-urban-household-energy-use-patterns-in-India_ACEEE-2020.pdf, 17 June 2021.

UN Department of Department of Economic and Social Affairs, Population Division (2017). Household Size and Composition Around the World 2017—Data Booklet. Retrieved from www.un.org/development/desa/pd/content/household-size-and-composition-around-world-2017-data-booklet, 23 June 2021.

The US Bureau of Labor Statistics (2021). Consumer Expenditure Surveys. Retrieved from New York - 2018 Level Expenditure Income Tables (bls.gov), 14 June 2021.

United States Census Bureau (2021). Computer and Internet Use in the United States: 2018. Retrieved from www.census.gov/newsroom/press-releases/2021/computer-internet-use.html#:~:text=Among%20all%20households%20in%202018,had%20a%20broadband%20internet%20subscription, 14 June 2021.

The US Library of Congress (n.d.) Family Life. Retrieved from http://countrystudies.us/north-korea/34.htm#:~:text=Both%20in%20urban%20areas%20and,oldest%2C%20son%20and%20his%20wife, 23 June 2021.

Wackernagel, M. and Rees, W. (1996). *Our Ecological Footprint: Reducing Human Impact on the Earth*. Gabriola Island, BC: New Society Publishers.

The World Bank (n.d.). World Bank Country and Lending Groups. Retrieved from https://datahelpdesk.worldbank.org/knowledgebase/articles/906519#High_income, 9 June 2021.

The World Bank (2021a). GNI per capita, Atlas method (current US$). Retrieved from https://data.worldbank.org/indicator/NY.GNP.PCAP.CD, 9 June 2021.

The World Bank (2021b). Access to electricity (% of population)—Kenya. Retrieved from https://data.worldbank.org/indicator/EG.ELC.ACCS.ZS?locations=KE, 21 June 2021.

World Economic Forum (2018). Future of Consumption in Fast-Growth Consumer Markets: India. Retrieved from www3.weforum.org/docs/WEF_Future_of_Consumption_Fast-Growth_Consumers_markets_India_report_2019.pdf, 16 June 2021.

Xpatulator (n.d.). Korea Democratic Republic of, Pyongyang. Retrieved from www.xpatulator.com/cost-of-living-review/Korea-Democratic-Republic-of-Pyongyang_233.cfm, 22 June 2021.

Yoon, L. (2021). North Korea in Crisis: Food Shortages and Information Lockdown. Retrieved from www.hrw.org/news/2021/03/16/north-korea-crisis-food-shortages-and-information-lockdown, 23 June 2021.

Yoon, Y.T. (2021). Comparison of the gross national income (GNI) per capita between South Korea and North Korea from 2010 to 2019. Retrieved from www.statista.com/statistics/1035411/south-korea-gni-per-capita-comparison-with-north-korea/#:~:text=GNI%20per%20capita%20comparison%20between%20South%20Korea%20and%20North%20Korea%202010%2D2019&text=In%202019%2C%20South%20Korea's%20GNI,1.41%20million%20South%20Korean%20won, 24 June 2021.

7

THE BUILT ENVIRONMENT

7.1 Introduction

The 19 cities have similarities in built form combined with differences that come from the local geography, climate and history. Traditionally cities tended to have a dense centre and a less dense periphery, the former often being contained by a wall as a form of defence. The suburbs, or the less dense areas, refer to the areas outside the wall, and by inference were seen as more vulnerable, with those living there retreating inside the walls in times of attack. Whitehand (1994) noted "We are living in built environments many of which have been bequeathed to us by our ancestors and which we shall in turn pass on to our descendants." Given this, the original intention of this investigation into the built form of the 19 cities was to group the cities historically, to gauge whether the history of the city has an effect on its modern built form and the impact of this on the environment. In Table 1.1, the cities were grouped into ancient cities (Beijing, Pyongyang, Cairo), medieval cities (Copenhagen, Moscow, Amsterdam, Hamburg, Singapore, Ahmedabad), early colonial cities (Havana, Sao Paulo, Buenos Aires, Jakarta, New York), and late colonial cities (Sydney, Dubai, Auckland, Kuala Lumpur, Nairobi). However, the difficulty in finding data on how the land is used in the cities meant that for most data was very limited, old or simply did not exist in an accessible form. For example, Table 7.1 shows the historic breakdown of land use in New York in 1995 calculated from data from the City of New York Department of City Planning.

The land area of New York City in 2019 was 302.64 square miles (783.8 km^2) (United States Census Bureau, n.d.), less than given in 1995; while the population in 1995 was 16.9 million, by 2019 it had risen to 18.8 million (Macrotrends, 2021). Obviously, these people have to live somewhere, but quite where and in what type of housing was not possible to discover. Detailed but outdated data for 1994 also exists for Nairobi (Oyugi and K'Akumu, 2007) (Table 7.2).

DOI: 10.4324/9781003045847-8

TABLE 7.1 1995 Land-use breakdown, New York (Demographia, 2000)

Land use	*Square kilometres*	*% total*
One-and-two family dwellings	181.5	22.9
Multi-family and mixed residential/commercial	83.8	10.6
Commercial	29.0	3.7
Industrial	55.6	7.0
Public facilities and institutions	48.1	6.1
Open space and outdoor recreation	127.8	16.1
Vacant land	31.3	9.8
Other	11.7	1.5
Streets	177.8	22.4
Total land area	792.6	100.0

TABLE 7.2 1994 land uses Nairobi

Land use type	*Area (ha)*	*% total*
High-income residential	11,000	16.1
Medium-income residential	4,070	5.9
Low-income residential	4,500	6.6
Unplanned slums and squatter settlements	2,190	3.2
Commercial	270	0.4
Industrial	2,410	3.5
Institutional	7,480	11.0
Recreational	15,330	22.4
Agricultural	21,150	30.9
Total	68,400	100

At the end of 2016, African City Planner (n.d.) states that former inner city residential areas were becoming mixed use and commercial, noting a change from single-unit dwellings that were becoming "multiple unit and eventually high density residential and, in some cases, commercial properties". A recent claim is that 60% of the Nairobi population lives in slums that occupy 6% of the city's area (Nairobi Slums project, 2021). The same source states a typical shack measures 12 ft × 12 ft (3.6 m × 3.6 m) made of "mud walls, a corrugated tin roof with a dirt or concrete floor" and might house up to eight people. The change from 3.2% of the city area in slum dwellings to 6% indicates the need for much more current data to make any worthwhile analysis.

In an age of land satellite images and geographical information systems (GIS) mapping, this seems a hole that could be filled, so that authorities could track changes in land use. Without such data, the aim here was to find an example of each type of city (ancient, medieval and colonial) and with judicious assumptions work out how much energy is embodied in its built environment. This calculation only deals with the energy to construct the various types of built environment, not with the energy necessary to maintain it, although this is reviewed in the discussion at the end of the chapter. Doing this proved possible to an extent, although results are for guidance only. This is an area worthy of more detailed exploration.

7.2 Ancient cities

Of the three ancient cities, there was no information on Pyongyang or Cairo, although a 2015 mapping exercise found the area of green space per person in Cairo was a very low 1.65 m^2 (Studio Meem, 2015). Russo and Cirella (2018) state that research on urban green space "points to at least 9 m^2 of green space per individual with an ideal UGS [urban green space] value of 50 m^2 per capita". Some information existed for land use in the central area of Beijing and this is explored below.

7.2.1 Beijing

The history of the development of Beijing has been well documented in terms of its stemming from the idea of duality, or *yin* and *yang* (female and male). As noted by Peng (1972):

> The theory of building a city or house was based on Confucian ideals—formality, symmetry, clarity, straight lines, regular forms, a strict order of human creation. The approach of designing a park or garden was guided by Taoist conceptions—informality, asymmetry, mystery, curved lines, irregular or romantic forms, a deep and persistent feeling for wild nature.

The walls of the Forbidden City are the start of a series of walls or enclosures that still permeate the city from the scale of the multiple ring roads, to the smaller preserved courtyard houses (Siheyuan) in the *hutongs* that have now become a tourist attraction (Travel China Guide, 2021). History still has an effect on the perceived value of parts of the city. "Referring back to the Imperial City layout, the south was still (or again) considered a 'poor people's' living place despite growing numbers of new and costly residential living complexes" (Fleischer, 2010:123–124) and to live in the north of the city gave an added prestige to those who could do this.

In 2000 in the absence of formal statistics, the estimates for housing types in Beijing were 25–30% high-rise (7 or more stories served by an elevator), 55–60% multi-storey (4–7 storey walk-up apartments) and 10–15% courtyard or 2–3 storey apartment buildings (Zhang and Brown, 2005). A more recent estimate is that more than 80% of all Chinese households live in high-rise and medium-rise apartments (Khan and Zheng, 2016:115). In 2016 the average urban per capita floor area was 36.6m^2 according to the National Bureau of Statistics (Xinhua, 2017).

The recent development of Chinese cities has tended to see an increase in green space as a percentage of total area rather than a decrease; this urban green space tends to be artificial and highly cultivated rather than natural (Zhao et al., 2010). In an analysis of green space in Chinese cities, green space in Beijing had risen from 20.69% in 2000 (713.31 km^2) to 34.35% in 2010 (12,25.47 km^2) (Yang et al., 2014). However, the original vegetation was not necessarily preserved.

> Expansion of BUAs [built-up areas] in Beijing has removed a significant percentage of original vegetation cover but the later efforts to increase UGC [urban ground cover] at any price have led to a final net increase of urban green spaces.
>
> *Yang et al., 2014*

Wu et al. (2020) give the following percentage breakdown of land areas from their study of Beijing (Table 7.3) based on 2012 data. Their study focussed "on the urban area within the fifth (from centre) ring road in Beijing, which covers most of the central urban districts that have the highest rates of urbanisation and greatest population densities." The area within the fifth ring road is 652.2 km^2 (Wang et al., 2013). The areas have been added to Table 7.3.

These figures only apply to the urbanised centre of Beijing, but the whole municipality is much bigger. As reported in 2015 (Anon): "More than half of Beijing's 21.5 million residents live outside the Fifth Ring Road."

The areas defined in Table 7.3 also contain land occupied by transport. According to the Beijing Municipal Commission of Transport, there were 29,400 km of road in Beijing by the end of 2018, with a road density of 179.3 km/km^2 (Xinhua, 2019). However, this road density figure does not seem feasible. NationMaster (2021) gives road densities in km/100 km^2 and the figure for Singapore (2007) as an urbanised nation is 471.68 km/100 km^2, or 4.7 km/km^2. Long and Zhang (2021:131) give road densities for the central areas of selected cities as follows: downtown St. Louis 6.22 km/km^2, Soho, London, 5.98 km/km^2 (although only 4.12 km/km^2 on p. 132 of the same source), HafenCity, Berlin, 5.34 km/km^2 and Old City Beijing,

TABLE 7.3 Land area breakdown for Beijing (2012)

Category	*%*	*Area (km^2)*
High-density residential	32	209
Commercial	16	104
Recreational	15	98
Industrial	15	98
Government	9	59
Development area	6	39
Public service	5	33
Agriculture	1	6
Low-density residential	0.5	3
Preservation area	0.5	3
Total	100	652

1.35 km/km^2 (although 1.77 km/km^2 on p. 132). This suggests the road density for Beijing should be taken as 179.3 km/100 km^2 or 1.8 km/km^2.

In the 1999–2014 extension of the city, the average road width was 7.3 m and roads occupied 26% of the expansion area (Atlas of Urban Expansion, 2016). However, applying the average width of roads to the length by the end of 2018 would give an area covered by roads of 215 km^2, which would only be 5% of the total area of urban Beijing of 4,172 km^2 (Table 1.3). However, some roads in Beijing are very much wider than 7.3 m. Jacobson et al. (2015) give the total 2013 road area as 279 km^2, which was 6.7% of the 2009 total area in the Beijing Municipality, which they gave as 4,139 km^2. They also note 7.2 m as the width of a residential street, while a motorway would be 28.7 m wide. Road coverage of 6.7% is lower than the 12–15% of metropolitan cities like London and Tokyo (Harmadi et al., 2015). However, these Beijing figures are quite old and despite policies to restrict car ownership, this is still predicted to rise, even under what is known as the tight constraints scenario (Li and Jones, 2016), and as car ownership rises there is more pressure to build roads. The UNHabitat (2013:106) study gives the proportion of land occupied by streets in Beijing as 19.1% in the city core and 6.1% in the suburbs with an average for the city of 12.6%, although it is not clear where the boundaries of the city were drawn.

As a further comparison, Whitelegg (1994:6) cites an early study in Germany that came up with a figure of 200 m^2 of asphalt and tarmac per vehicle, which would account for both roads and carparks, and quotes a Swiss study figure of 113 m^2 per person for road transport and 20–25 m^2 per person for houses, gardens and yards (Whitelegg, 1994:4). In 2018 "Beijing had 5.97 million locally registered cars as of April, according to the Beijing Commission of Transport. However, there are constantly more than 700,000 non-Beijing-registered cars in use in the downtown areas" (Du, 2018). Using the figures from the various sources gives the following area for roads in Beijing, based on 6 million vehicles (Table 7.4).

The figure of 1200 km^2 includes car parks and having somewhere to park your car in China can be more expensive than buying the vehicle (Guilford, 2014). In luxury apartments, cars will be parked in the basement, adding to the overall cost of the building. As the UNHabitat figure for the core was published in 2013, the

TABLE 7.4 Estimates of land cover by roads in Beijing

Source	*Road area (km^2)*
Whitelegg (German)	1200
Whitelegg (Swiss)*	2189
UNHabitat total**	526
Jacobson et al.	279

* Population 19,372,000 from Table 1.3.
** Area taken as 4,172 km^2

figure of 19.1% of total will be used in this study of the area within the fifth ring road (652 km^2) although this may be an overestimate.

Beijing has both overground railways and a subway system. DVV Media International (2019) notes: "The opening of two extensions to the Beijing metro on December 28 [2019] has taken the total network length to 699.3 km with 405 stations, including 62 interchanges, overtaking Shanghai to become the world's largest metro network." However, things move fast in China and two years later Song (2021) states there are 30 lines, 391 subway stations and 1,177 kilometres of railway line, although Shanghai is now the world's largest subway system. Most of the Beijing track is underground though there is some on-ground travel and also some elevated track on the newer lines. However, in the absence of further data, the assumption is that only half of this (588 km) lies within the fifth ring road area (see Figure 9, Gao and Newman (2018)), and it would all be underground.

The length of the four suburban railways in Beijing totals 400 km (Anon, 2021), of which just over 9 km run in a tunnel connecting two main railway stations (Anon, 2018). The width of land required for a double track is approximately 15 m (Connor, 2017). Assuming a track width of 15 m as the suburban trains run on standard gauge, the land area occupied by the suburban railways for the whole of Metropolitan Beijing would be 600 ha or 6 km^2, most of which will lie outside the central area being considered here.

Using 19.1% of total for the area of roads and putting these figures together, while accepting that not all are for the same date, gives Table 7.5 for the land allocation for the area within the fifth ring road.

TABLE 7.5 Breakdown of land cover in Beijing within the fifth ring road accounting for road area

Land cover type	*% total*	*Area (km^2)*
Roads	19.1	125
High-density residential (32%)	25.9	169
Commercial (16%)	12.9	84
Recreational (15%)	12.1	79
Industrial (15%)	12.1	79
Government (9%)	7.3	48
Development area (6%)	4.8	32
Public service (5%)	4.0	26
Agriculture (1%)	0.8	5
Low-density residential (0.5%)	0.4	2.5
Preservation area (0.5%)	0.4	2.5
Total	99.8*	652

* Numbers rounded.

7.2.2 Embodied energy of Beijing buildings and infrastructure within the fifth ring road

In order to see the comparative impact of this built environment only in terms of the materials of which it is made, a simplified embodied energy calculation is used and only the initial embodied energy—the energy it takes to make a building or road—is considered. As the energy embodied in the built environment changes with the type of energy used to make the buildings and infrastructure, figures for Beijing are drawn from available Chinese sources.

In a study of six Beijing office buildings, Li et al. (2019) calculated an embodied energy figure of 7.73 GJ/m^2 while noting values of 9.74 and 6.91 GJ/m^2 for similar buildings from other sources. Their six case study buildings comprised three of ten storeys and three of eight storeys, both with two basement car parking levels. The construction area of the six buildings was 4,640 m^2 (0.00464 km^2) and the total energy embodied in these buildings was 540 TJ (not 540 PJ as stated) for a total built floor area of 69,844 m^2, but it is not clear whether construction area equates to the footprints of the buildings or is the whole building site area. Pan et al. (2020) note that for the whole of China the average floor area per employee in the commercial sector was 32.7 m^2. In 2019 83.1% of workers were involved in the tertiary sector in Beijing of 12,730,000 workers (Beijing Municipal Bureau of Statistics, 2020:table 3-11). Very approximately, assuming 12 million workers, 30 m^2/worker and an embodied energy of 7.7 GJ/m^2 would give an embodied energy (EE) of 2,772,000 TJ or 2772 PJ for all tertiary sector employees, thus covering anyone not working in acquiring raw materials (the primary sector), such as mining or agriculture, or in the manufacturing or assembly industries (the secondary sector), which would include construction.

In a study of three steel frame residential buildings in Dujiangyan, Sichuan province in south-west China, Su and Zhang (2016) found little correlation between the size of the buildings and the embodied energy (Table 7.6). The norm for residential buildings in China is reinforced concrete and a steel frame was used here because of being in an earthquake zone.

In a study of an academic building in Shijiazhuang, Hebei Province in north China (lying south of Beijing), Chang et al. (2012) also calculated an embodied energy of 6.3 GJ/m^2. This building was 19 storeys with a podium and two basement parking levels. Both these studies included transport (bringing materials to site) and construction (making the building on site) energy.

TABLE 7.6 Embodied energy of steel frame residential buildings (Su and Zhang, 2016)

Building area (m^2)	*No. of floors*	*EE (GJ/m^2)*
5524	6	6.3
5544	11	6.3
9574	18	6.7

An Australian study of reinforced and post-tensioned reinforced concrete residential buildings (Miller et al., 2014) found values of 7.84 GJ/m^2 (20 floors, 2 basements), 6.78 GJ/m^2 (4 floors, 1 basement), 7.5 GJ/m^2 (6 floors, no basement) and 6.77GJ/m^2 (3 floors, and 1 basement). For another comparison, an Indian study, which included the energy in the human labour component of construction (Abey and Anand, 2019), which they stated can range from 1% to 14% of total EE, gives a figure of 4.02 GJ/m^2 for a reinforced concrete apartment of four floors, with four apartments per floor. From all this, a figure of 6.3 GJ/m^2 (the value from Chang et al.) for high-density residential seems reasonable. The problem comes in knowing the residential floor area within the fifth ring road. Nie et al. (2022) note that those living within the fifth ring road "are well-to-do middle class". In the absence of data, the assumption is that slightly less than half of all Beijing's inhabitants live within the fifth ring road (10 million people) and the per capita living space is 32.5 m^2 (the value for 2019 for urban Beijing from Slotta (2021)). This gives an embodied energy for residential buildings of 2,047,500 TJ or 2048 PJ and an EE intensity of 12.2 PJ/km^2. As a check, this value can be used to reassess the energy embodied in the 132 km of commercial and government buildings using a value of 7.7 GJ/m^2. Given commercial buildings are high rise and the area per employee is very similar to the residential area per person, the proportional energy intensity of commercial buildings would be 14.1 PJ/km^2, and the 158 km^2 of this building type (commercial, government and public service) would have an energy intensity of 2,228 PJ. This is of the same order as the 2,772 PJ for the tertiary sector in Beijing calculated above.

The only data found for an industrial building was from a study in Portugal by Rodrigues et al. (2018). The two-storey building occupied 27% of the total site, had an embodied energy of 4.9 GJ/m^2 and "6733 m^2 of effective construction area". This gives a value of 32,992 GJ (33 TJ) for the building on its total site area of 21,400 m^2 (0.02 km^2). From Table 7.5, there are 79 km^2 of industrial buildings, which pro rata would have an embodied energy of 130,350 TJ (130 PJ). The Portuguese study gives a view of the site showing that the area not occupied by the building is hard surface used for parking vehicles and containers. To have a sense of the importance of this area in embodied energy terms, a study in India concluded that the average energy associated with asphalt and cement concrete pavements varied from 700 to 1,000 MJ/m^2 (Singh et al., 2020). Since 73% of the area of the study site is not covered by buildings, this would give a range of 41–58 PJ assuming all the areas around industrial buildings are hard landscape, which may be an overestimate.

For the Beijing Subway, it would appear that most will be in two single-track tunnels, as Li et al. (2016) note in reference to the new line linking the Dongfengbeiqiao and Jingshunlu areas of the city that 2.8 km of a double track tunnel would be used because of space limitation. In a study of the twin-bored UK Channel Tunnel link, Chau et al. (2012) found a 7.5 km section of twin-bored tunnel had an embodied energy of 949 TJ whereas [7.3] km of a mostly pile-slab structure would require 678 TJ. Assuming therefore a figure of 100 TJ/km for tunnels would give 58,800 TJ (58.8 PJ) for the embodied energy of the tunnel component of the Beijing Subway system.

Mithraratne (2011) gives a value of 8,506 GJ/km for rail track in New Zealand. Applying this to the 588 km of route with 1,176 km of total track would give an embodied energy of 10,003 TJ (10.0 PJ). This would give a total of 68.8 PJ for the subway within the fifth ring road.

Mithraratne (2011) gives a value of 646 GJ/lane kilometre for road pavement on a flat site. In their study of traffic flow and lane width in Beijing, Chang et al. (2019) found lane widths of 2.6–3.5m. In Europe for temporary lanes due to road works lane widths of 3.25–2.5 m are recommended depending on the speed and direction of traffic (EuroTest, n.d.). Knowing the roads cover an area of 125 km^2, and assuming an average lane width of 3 m gives an embodied energy of 26,917 TJ (26.9 PJ). The Mithraratne figure is for New Zealand and seems to fall at the low end of possible embodied energy.

> A meta-study at the University of Washington found that a reasonable estimate of total energy consumption is between two and four terajoules per lane-km of roadway paved. A lane-km is defined as a kilometer long segment of roadway that is a single lane in width (for example, a one kilometer stretch of a standard two lane road represents two lane-km). The lowest values noted in this study found values less than a single terajoule per lane-km, while the highest values exceeded seventeen terajoules per lane-km.
>
> *Pavement Interactive, n.d.*

The energy that goes into making a road is affected by the amount of rock and earth moved, which can give rise to the very high values of embodied energy (Karlsson et al., 2017). As Beijing is a flat city, this is much less relevant. Using the highest value of 4TJ/lane kilometre and again assuming a lane width of 3 m would give a more significant embodied energy value of 167 PJ for roads within the fifth ring road.

The areas not yet tackled from Table 7.5 are recreational (79 km^2), development (32 km^2), agriculture (5km^2), low-rise residential (2.5 km^2) and preservation area (2.5 km^2). There are significant preserved areas within the second ring road in the city centre, including the Forbidden City and Shichahai, which together cover an area of 3.7 km^2 (Zhang, 2008). Apart from ancient monuments and a lake, Shichahai also contains historic low-rise courtyard houses. Given the age of these monuments and houses and the small area they cover, their historic embodied energy is ignored. The development area is assumed to be cleared land and any buildings associated with the small area of agriculture can also be ignored. The problem with recreational space is that it could include some built structures. In their study of open space use in central Beijing, Li et al. (2017) identified an area of 65.85 km^2 of urban green spaces, a small part of which would be within the preservation area discussed above. However, this included 22.72 km^2 of theme park and areas like the Aquatic

TABLE 7.7 Energy embodied in Beijing within the fifth ring road

Category	*Embodied energy (EE) (PJ)*	*EE/person* (GJ)*	*EE density (PJ/km²)*
Commercial/government/public service	2228–2772	223–277	14.1–17.5
High density residential	2048	205	12.1
Industrial	171–186	17–19	2.2–2.4
Roads	27–167	2–17	0.2–1.3
Subway	69	7	0.1**
Total***	4893	489	

* Based on an assumed 10 million people.
** Using the whole area of 625km² within the fifth ring road.
*** Using the midpoint of the ranges.

Park that was part of the Beijing Olympics. Buildings on such land could already be accounted for under the commercial category. Looked at another way, 79 km² of recreational area would give 7.9 m²/person assuming 10 million people live within the fifth ring road. Assuming this is green space of some kind this is low compared to other cities (Washington 38 m²/person; Edinburgh 30m²/person; Bristol 10m²/person (Khan, 2014)). Area per person is also a very crude measure that does not take account of green space accessibility (WHO, 2016). Given this, the assumption here is that the land designated as recreation is green public space crossed by paths and other access ways, the embodied energy of which is ignored at this stage.

In order to see which elements of the built environment contribute most to the energy embodied in it, Table 7.7 sets out the figures calculated above, noting that these are based on significant assumptions and represent only the part of Beijing within the fifth ring road.

Table 7.7 shows the big energy is embodied in buildings rather than infrastructure. The industrial building figure is lower because the buildings are assumed to occupy less than a quarter of the site with the remainder being hard standing. The range for roads is great and more investigation into how the roads are constructed is necessary. The subway, which serves many more people than the roads in terms of passenger-km travelled is good value in terms of the energy required to construct it.

Assuming a 50-year life for everything, and ignoring the energy that goes into maintenance and upkeep of buildings and infrastructure, Table 7.7 gives an annual energy consumption for embodied energy of the built environment of 9.1–10.5 GJ/person, which is roughly three times the energy used for transport per person.

7.3 Medieval cities

Of the medieval cities, data was found for Amsterdam, Ahmedabad and Singapore and these are discussed below.

7.3.1 Amsterdam

Table 7.8 sets out the breakdown of land use in Amsterdam for 2012 (OECD, 2017:61) (numbers rounded), although it is not clear exactly where the boundaries of the city have been drawn.

The final column needs some explanation. Amsterdam is known for its canals, an important part of the drainage of a city that lies just below sea level. It appears a city still dominated by water. "Almost half of the original water in Amsterdam was lost to landfills, but a full 25 percent of the city's surface still consists of navigable waterways. With 65 miles [over 100 kilometres] of ancient canals, Amsterdam is still the most watery city in the world" (Amsterdam.info, 2021). This area of water is considerably less than the OECD figure for 2012 (Table 7.8) but noticeably refers to navigable waterways, not all water, which would include the lakes of Amsterdam. Netherlands Insiders (2021) make the point that "The number of canals varies over time because sometimes a canal is turned into a road and vice versa." They give the total length of canals as 65 km. However, for a fairer comparison it seemed better to use the percentages in the final column, which deals only with the built environment. The area for recreation is, as in the case of Beijing, assumed to have zero embodied energy.

Transport only accounting for 5.2% of total area, which is very low (Harmadi et al., 2015), needs some explanation. Cycling is very important and despite the growth in car ownership and the presence of motorways, Amsterdam is one of the few cities in the world that can lay claim to being a cycling city. The modal split for all journeys is walking 31%, cycling 32%, public transport 17% and motorised private vehicles 20% (Kodukula et al., 2018:13), so public transport use is less and cycling greater than in other European cities, apart from Copenhagen where cycling is also important and where the modal split is walking 19%, cycling 29%, public transport 18% and motorised private vehicles 34%. Amsterdam has a historic core of narrow streets and canals and as long ago as 1927, an Amsterdam city councillor noting the rise in traffic stated that "either the existing city should be adapted to the needs of modern traffic, by breaking down houses and filling in canals, or the new traffic should adapt to the possibilities of the old city" (Feddes et al., 2020:151). The

TABLE 7.8 Breakdown of land use in Amsterdam

Land use	*Hectares*	*% total*	*% land area only*
Inland waters	5401	44	
Residential and commercial	4463	36	64.5
Semi-built-up areas	1165	9	16.8
Recreational	709	6	10.2
Transport	362	3	5.2
Forest and open terrain	194	2	2.8
Agricultural	30	0.2	0.4

same authors who reproduced this quote give a detailed account of how Amsterdam managed to maintain cycling as an important mode of transport, not least through the efforts of ordinary people to influence council policy.

Land take for cycling is also much lower than for cars. An early German study comparing the land take for passenger travel (Whitelegg, 1994:1–2), apportioned 120 m^2 per car used for travelling, 12 m^2 for bus, 9 m^2 for a bike, 7 m^2 for a journey by train, and 2 m^2 for walking. When it came to freight, a lorry needed three times the space per tonne-kilometre travelled than carriage by train. Looked at another way, a 3.5 m wide city road could carry 2,000 persons/hour in cars in normal mixed traffic, 9,000 in buses, 14,000 cyclists and 19,000 pedestrians (Hickman et al., 2011:55). Given the modal split for Amsterdam above this helps to explain the small area of the city set aside for transport. Amsterdam also has 200 km of tram tracks embedded in the roads (GVB, 2013). GVB (Gemeente Vervoerbedrijf) operates public transport in metropolitan Amsterdam.

There are 120 km of rail tracks for the Metro system, which serves the city and the surrounding areas (GVB, 2013). It appears that where the metro lines pass through the old city area they are underground, some in bored tunnels at depth and others just under streets, being built with more than one technology (Van Lohuizen, 1989). The central area of Amsterdam occupies 8 km^2 including the canals (Onderzoek, Informatie en Statistiek, 2019:10). The total area of the city from Table 7.8 is 123 km^2 so the centre accounts for 6.5% of the area. Since from the map of the Metro (AmsterdamTips, 2021), the majority of the lines are outside the central area the assumption here is that 6.5% of the 120 km of track (7.8km) is in tunnels of some kind. Using the width of 15 m gives a land area of 1,800,000 m^2 or 1.8 km^2. The total area occupied by transport from Table 7.8 is 362 ha (3.6 km^2), so the area occupied by roads and cycle ways is also 1.8 km^2.

According to CBS (2021), in 2015 dwellings in Amsterdam were smaller than in the rest of the Netherlands, being an average 74 m^2 per household. Apartment dwellers formed 85.8% of the population, with 13.1% in terraced housing, 0.4% in semi-detached houses and 0.6% in detached houses. In 2015, the average household size in the Netherlands was 2.17 persons (34 m^2/person). This compares with an earlier figure of 32 m^2/person for the whole country from Statistics Netherlands cited in de Wilde and van den Dobbelsteen (2004). Given the very approximate nature of these calculations, the average of 74 m^2 per household is used here. From Table 1.3, the population of Amsterdam is approximately 1.1 million, giving 506,912 households of which 434,930 live in apartments, 66,405 live in terraced houses and 2,028 in semi-detached and 3,041 in detached houses.

In 2020 "approximately 715,000 square meters of lettable floor area was available in the region of Amsterdam" (Statista Research Department, 2021), although this only accounted for parcels of lettable floor area of 500 m^2 or more. However, this does not cover other aspects of commercial buildings, such as retail, and no account has been taken of other large buildings such as hospitals and schools.

7.3.1 Embodied energy in built Amsterdam

Koezjakov et al. (2018) found the embodied energy of Dutch housing types varied from 3 to 6.4 GJ/m^2 with apartments having lower values than mid-terraced houses because of reinforcing the party walls. In the absence of further data, a value of 3 GJ/m^2 had been assigned to apartments and 6GJ/m^2 to all other housing types. The embodied energy results are shown in Table 7.9.

The total energy embodied in Amsterdam's housing is thus 128,468 TJ (128PJ).

In 2000 de Wilde and van den Dobbelsteen (2004) gave an area of 24 m^2 for each office employee and 48 m^2 per person for amenities, the latter covering all buildings apart from housing and offices. In the absence of more recent data, these values have been used here. In 2018 out of a population of 674,000, 68% were in work (458,000), 3% were unemployed (23,000) and 29% were not working (193,000) (Onderzoek, Informatie en Statistiek 2019:113). Assuming an average area per employee of 36 m^{2}and applying 7.7 GJ/m^2 for commercial buildings (Li et al., 2019) gives an embodied energy of 186,833TJ (187PJ).

The embodied energy of 1.8 km^2 of railways using the same figure as for Beijing (Mithraratne, 2011) of 8,506 GJ/km is 1,021 TJ (1.0 PJ). Using the same data of 100 TJ/km, the energy embodied in the 7.8 km of tunnel is 780TJ (0.8PJ).

The area occupied by roads and cycleways is 1.8 km^2. Assuming an average lane width of 3 m and, in the absence of more local data, a value of 4 TJ/lane kilometre, as for Beijing, gives an embodied energy of 2,400 TJ (2.4PJ). This is probably an overestimate for the cycleways component, which may well have a different construction from roads. However, as the steel tram tracks embedded in roads have not been calculated, this figure has been used in Table 7.10.

TABLE 7.9 Residential embodied energy, Amsterdam

Housing type	*Area (m^2)*	*Embodied energy (TJ)*
Apartments	32,184,820	96,554
Terraced houses	4,913,970	29,484
Semi-detached	150,072	900
Detached	225,034	1,530

TABLE 7.10 Energy embodied in Amsterdam's built environment

Category	*Embodied energy (EE) (PJ)*	*EE/person* (GJ)*
Residential	128	190
Other buildings	187	277
Roads	2	3
Metro	2	3
Total	319	473

* Based on population 674,000.

7.3.2 Ahmedabad

As a comparison with a different type of medieval city, Ahmedabad was also investigated. Because the method and the type of assumptions that need to be made have been explained in the Beijing and Amsterdam examples, here the categories and embodied energy are considered together.

Table 7.11 sets out the land-use breakdown from the 2011 census, the most recent available (Annez et al., 2012:16). These areas would include roads and open space. A gamtal is a village area that has been surrounded by the city and that may not have urban infrastructure. Challis or chawls are very substandard housing, traditionally provided by mill owners for their workers on private land.

The same authors give a percentage for road cover in new town planning areas of 11.69% of developed area (Annez et al., 2012:27). Bharti and Mehrotra (2020:18) state: "The city has a good network of roads totaling 2,399 kilometers." The same authors also note that Ahmedabad, unlike many cities in the world, has avoided serious urban sprawl by increasing density (Bharti and Mehrotra, 2020:5). This suggests that since the 2011 census housing will have intensified. Further evidence for this comes from the fact the open space area per person is also very low at 0.37 m^2/person against a standard of 8–10 m^2/person (Patel 2016:50). The Ministry for Urban Development (n.d.) describes how Ahmedabad grew outwards with a series of circular ring roads around the old city crossed by radial roads. In 2011 "the share of area under the roads constitutes 7.5% of the entire city area. Within the developed area, the area under roads is about 18%. This translates to an average road width of 12 m and about 10 kms per sq.km of developed area." Outside of the developed areas and within the slums and gamtals, the roads will be narrow and possibly unpaved. Taking the length of 2,399 km and an average width of 12 m gives an area of roads of 28.8 km^2. Accepting that the total area is for the 2011 census, while the length of roads applies to 2020, this would give a road area of 13.6% of the total, and an adjusted area of 183.38 km^2 for all other land uses. The roads in the slums and challis will not be to a high standard so this total area has been maintained with no reduction and all other areas adjusted on a pro rata basis.

TABLE 7.11 Ahmedabad land use breakdown

Land use	*km^2*	*%*	*Adjusted areas to take account of roads (km^2)*
Mixed built-up	142.51	67.1	107
Old city	5.60	2.6	6
Slums and challis	15.29	7.2	15
Gamtals	11.69	5.5	13
Industry	16.69	7.9	19
Villas in large estates	20.47	9.6	23
Total	212.25	99.9 (numbers rounded)	183

In their study of Indian building materials, Venkatarama Reddy and Jagadish (2003) calculated a value of 4.2 GJ/m^2 for a multi-storey reinforced concrete frame building. This compares with the more recent 4.02 GJ/m^2 of Abey and Anand (2019) for a reinforced concrete apartment building. The floor area of each storey of this four-storey building with four two-bedroom apartments on each floor was 232 m^2 and allowing for circulation space each apartment was approximately 50 m^2. The energy embodied in the four-storey building was thus 3,731 GJ (3.7 TJ).

Annez et al. (2012:12) state the vernacular housing stock in Ahmedabad covers 84% of the site and is three floors. They also note (2012:14) that in 2011, 3.7 million people lived in areas of formal housing occupying 197 km^2, which includes all other categories in Table 7.11, while 1.9 million people lived in slums and challis occupying 15.3 km^2. Applying the figure for the four-storey building to the three storeys of Ahmedabad gives a value of 2,798 GJ (2.8TJ) for a residential building with a footprint of 232 m^2. If this same building covers 84% of the site, this would give a residential embodied energy density of 10.1 GJ/m^2 of land. In Table 7.11, the adjusted area covered by mixed built-up allowing for roads was 107 km^2. Assuming all buildings both residential and commercial in this zone have the same embodied energy, the energy embodied in this part of Ahmedabad is 1,080 PJ. However, this does not account for the possibility of more energy embodied in commercial buildings. There are plans to build a new CBD with much higher buildings on reclaimed land along the river to "change the skyline of the city" (Sharma, 2021). Any increase in EE in the CBD has been ignored at this stage.

It has not been possible to explore the nature of the housing in the gamtals of Ahmedabad. Looking at a study of an urban gamtal in Surat, a city to the south of Ahmedabad but also in Gujarat, the area consisted of two- and three-storey masonry buildings alongside narrow streets (Udajea et al., 2020). Venkatarama Reddy and Jagadish (2003) give a value of 2.92 GJ/m^2 for a 150 m^2 two-storey load-bearing structure of brickwork walls, reinforced concrete roof slab and mosaic finish. Earlier, Debnath et al. (1995) gave 3 GJ/m^2 as the lowest figure for residential buildings, from their study of "45 single storey, 35 double storey load-bearing wall residential buildings, and 60 four storey framed structure residential buildings." It would seem reasonable, therefore, to apply a figure of 3 GJ/m^2 to housing in gamtals. The adjusted area to take account of roads is 13 km^2, which is also assumed to include any open space, as the roads are narrow and no allowance has been made for this. Assuming half the buildings are three-storey and half two-storey would give a total floor area of 33.5 km^2, with an embodied energy of 100,500 TJ (100.5 PJ).

The slums and challis are here assumed to be made of materials to hand that are probably recycled and so their embodied energy content could be assumed to be zero. It is also fair to say that this investigation of Ahmedabad is based on old data and the situation is changing since the 2010 launch of the slum-free India government initiative (Singh, 2010). Rather than displacing people, the aim was to rehouse people from the slum in the same place in apartment blocks provided free by developers.

> In exchange, the city would give the contractors any leftover land, which they could use to build expensive apartments for sale on the open market. The city also increased the "buildable area" allowances, meaning taller, denser buildings—a very lucrative incentive for developers to get involved in housing for the poor.
>
> *Carr, 2016*

Such redevelopment will raise the embodied energy of the built part of the city but will also alleviate the problem of living in slum conditions. The materials embodied in the 6 km^2 of the old city have also been ignored, given the same was done for the Beijing historic area.

The Indian government states that Ahmedabad is an industrial centre with, in 2011, "industries related to textiles, chemicals, machinery, metal products, pharmaceutical, engineering, plastics, electrical appliances, electronics, passenger cars etc." in the city and the surrounding district (Ministry MSME, n.d.). In the absence of any data on embodied energy in industrial buildings in India, the values used for Beijing have been applied here (33 TJ/0.02km^2) to give a value of 31,350 TJ (31 PJ).

Ahmedabad has significant areas of low-rise houses set in their own plots. Looking at houses for sale on the web indicates these are mostly two-storey and some three-storey but with differing footprints depending on cost. According to the local regulations, the floor space index (FSI), which is the ratio of the permitted built area to the area of the plot, is 1.8 for residential zone one, which is the main residential zone of the city, though this can be increased through a charge to 2.7–2.25 depending on the location (AUDA, n.d.:57). An FSI of 1.8 for a two-storey house would mean the building footprint could occupy 90% of the plot area, though not all would perhaps do this. The area of villas in large estates is 23 km^2. Assuming 80% of this is covered with two-storey houses with an embodied energy of 4 GJ/m^2 to reflect the high build quality gives an embodied energy of 147,200 TJ (147 PJ).

A recent study of pavement construction using Indian embodied energy data (Singh et al., 2020) produced a range of 1,750–1,786 GJ/km depending on the type of construction, for a pavement 3.5 m in width. Taking this data and the 28.8 km^2 of roads in the city, this would have an embodied energy of 14,400 TJ (14.4 PJ) to 14.696 TJ (14.7 PJ).

Ahmedabad has a bus rapid transit system although there is a project for a new metro under construction of just over 40 km of railway line of which some 6 km will be underground (Verdict Media Ltd., 2021). As the land use breakdown is for 2011, the energy embodied in this has been ignored in Table 7.12. There is also a planned suburban railway system with 52.9 km of broad gauge that would use the existing rail corridors in the city. As these rail corridors exist, it seems reasonable to include their embodied energy. Using the figure of 8,506 GJ/km (Mithraratne, 2011) gives an embodied energy of 450 TJ (0.45 PJ). With roads, this gives a combined transport infrastructure embodied energy of 15.15 PJ. Table 7.12 sets out the embodied

TABLE 7.12 Energy embodied in Ahmedabad's buildings and infrastructure

Category	*EE (PJ)*	*EE/person (GJ)*	*EE density (PJ/km²)*
Mixed built-up	1080	194	10
Old city	-		-
Slums and challis	-		-
Gamtals	101	18	8
Industry	31	6	2
Villas	147	26	6
Transport infrastructure	15	3	0.5
Total	1374	247*	N/A

* Numbers rounded.

energy in the buildings and transport infrastructure of Ahmedabad. The EE per person is based on the 2011 population of 5,577,940 (Census Population, 2021).

7.3.3 *Singapore*

Singapore being a confined city-state with a high population not only has figures for reasonably current land usages but also projections for future land uses, to aid planning given population growth (Ministry of National Development, 2013). This report notes:

> To support this larger population, we need to (a) reclaim additional land; (b) develop some of our reserve land; (c) intensify new developments; and (d) recycle land with lower intensity uses such as old industrial areas and some golf courses to achieve higher land productivity.
>
> *Ministry of National Development, 2013:4–5*

Table 7.13 gives the most recent and 2030 projections for land use.

The difference in this table from other available city land usage is the 19% that Singapore gives to defence because it is a city-state, and this despite its use of overseas training facilities. "The city state's lack of space has led it to put in place an elaborate web of overseas training arrangements, involving a dozen countries. Australia is now emerging as the most important of these…" (Graham, 2021). This area of land has been ignored in the following assessment of embodied energy. As before, parks and nature reserves are assumed to have zero embodied energy. There is farmland in Singapore, but increasingly this has given way to development so the "Other" category is assumed to be land of this type as it has significantly reduced by 2030 and is assumed to have zero embodied energy. The airport is also ignored as not all cities considered have major airport hubs like that of Singapore. Similarly, the energy embodied in the reservoirs and utilities is ignored as it had been in the estimates for the other cities.

TABLE 7.13 Land use breakdown, Singapore

Land use	*2012 (ha)*	*2012 %*	*2030 (ha)*	*2030 %*
Housing	10,000	14	13,000	17
Industry and commerce	9,700	13	12,800	17
Parks and nature reserves	5,700	8	7,250	9
Community, institution and recreation facilities	5,400	8	5,500	7
Utilities (e.g. power, water treatment)	1,850	3	2,600	3
Reservoirs	3,700	5	3,700	5
Land transport infrastructure*	8,300	12	9,700	13
Ports and airports	2,200	3	4,400	6
Defence requirements	13,300	19	14,800	19
Other	10,000	14	2,800	4
Total	71,000**	100	76,600	100

* This seems to refer to roads as the same report states "Roads already account for a significant 12% of our total land use" (Ministry of National Development, 2013:51) but will also need to account for the MRT railway.

** This should be 70,150.

In a ranked list of Asian countries (Chew, 2019), Singapore had residential living space of 32.9 ft^2/person (30 m^2/person). From Table 1.3, the population is 5,745,000. There is an absence of data on residential buildings in Singapore. Applying the value of 6.3 GJ/m^2 for high-density residential buildings (Chang et al., 2012) would give an embodied energy of 1,085,805 TJ (1085 PJ).

A study of a commercial/industrial building in Singapore that occupied 27,029 m^2 with a floor area of 52,094 m^2 found the energy embodied in its materials was 27.9 TJ. This building was stated as being "exemplary for green design for multi-storied commercial buildings in Singapore" (Kua and Wong, 2012). However, this would give a very low 0.5 GJ/m^2, whereas 5 GJ/m^2 would be reasonable. Allowing for this error and applying the latter rate to the 9,700 ha of industrial and commercial buildings gives an embodied energy of 1,002,333 TJ (1,002 PJ).

Chang et al. (2019) studied the energy embodied in the National University of Singapore (NUS) buildings and found a range of 2.0–3.6GJ/m^2 but noting that these values were lower than those of other studies "due to the exclusion of other building materials, substructures and building foundation". The other problem is that the NUS campus has considerable open areas, which may well apply to other community and institutional facilities. NUS occupies 165 ha with a gross floor area of 1.5 million m^2 of which 24% are residences (NUS, 2019). This leaves 1,140,000 m^2 of non-residential built floor area. To allow for foundations but reflect the study of Chang et al. (2019), a value of 4 GJ/m^2 is applied to give an embodied energy of 4,560 TJ in 125 ha, allowing for the residences. Applying this to the 5,400 ha of community and institutional buildings would give an embodied energy of 196,992 TJ (197 PJ). However, not all institutional and community buildings will be set in such leafy surroundings as NUS. Singapore General Hospital, which is set in 43 ha (SingHealth Group, 2021), appears to have half the site covered by buildings, most of which are high rise. It therefore seems better to apply the commercial building embodied energy density of 103 TJ/ha to the 5,400 ha of community, institutional and recreational buildings to give an embodied energy of 556,200 TJ (556 PJ).

Wei (n.d.), in a study of development in the Jurong area, states the energy "to make 1 km of the road can be reduced from 0.93TJ to 0.37TJ annually" by using incinerator bottom ash in the place of gravel, although also noting this would put more heavy metals into the constructed roads. In 2014, the Singapore Land Transport Authority was responsible for 3,496 km of roads (Singapore Government Statistics, n.d.), which using 0.93 TJ/km would have an embodied energy of 3,251 TJ (3 PJ). As the 2019 road length is only marginally larger at 3,520 km (ASEANstats, 2018), this estimate is accepted.

The Singapore MRT railway is both underground and elevated. It has 200 km of double-tracked line and an additional 28 km of light rail line, which is elevated (Government of Singapore, 2021). In the absence of data and based on experience of the MRT, the assumption has been made that half of the 200 km is underground and half elevated, noting that the underground is deep in places as it also serves as a bomb shelter. Using the figure of 100 TJ/km used earlier for the tunnel part of the MRT gives an embodied energy value of 10,000 TJ (10 PJ). Using the value of 8,506 GJ/km for rail (Mithraratne, 2011) for the 456 km of track produces an embodied energy of 3,879 TJ (4 PJ). An embodied energy for bridges has been used for the 228 km of elevated line. Penadés-Plà et al. (2019) give values of 744,156–771,853 kWh for a 130 m long continuous box girder pedestrian bridge, or 21 TJ/km. Mithraratne (2011:20) notes "the carbon footprint due to construction of bridge deck per km is 7125tCO2e, or 37.5 times that of the road." The figure of 21 TJ/km is 22.5 times the value of the embodied energy used for roads above, and this value is also an underestimate as it does not account for all the supporting piers, only for the deck itself. In the absence of further data, 21 TJ/km has been applied to the 128 km of elevated track for a total embodied energy of 2,688 TJ (3 PJ). The total embodied energy of the MRT is thus 17 PJ.

Table 7.14 sets out the embodied energy of the measurable parts of the infrastructure of Singapore, using the population of 5,745,000 from Table 1.3.

TABLE 7.14 Embodied energy of part of Singapore's built environment

Category	*EE (PJ)*	*EE/person (GJ)*
Residential	1085	189
Commercial	1002	174
Community, institutional, recreational	556	97
Roads and MRT	20	3.5
Total	2663	463.5

7.4 Colonial cities

The only data available was for the city of Jakarta, which as Batavia was for three centuries under the control of the Dutch.

7.4.1 Jakarta

Herlambang et al. (2019) state "Locally, Jakarta's urban landscape is still redolent of the colonial and post-colonial processes that generated a city where the majority of residents live in kampungs located between the major thoroughfares." Earlier Lo (2010) noted that Jakarta grew with "commercial ribbon development along key roads and sprawling new towns linked by regional tollroads". The same author also stated that until the 1960s some 60% of all trips were walking, a situation that changed along with so many cities with the advent of the private car.

Although no land-use data was found for central Jakarta, there is information on three districts in the city, as set out in Table 7.15 (Putri et al., 2019).

Penjaringan lies to the north along Jakarta Bay and houses the port, which explains the area set aside for transportation. Cenkareng lies south of Penjaringan in west Jakarta and Cakung is east of the central area of the city. Road areas are included within the areas in Table 7.15. When it comes to housing, which would be included in the settlement category in Table 7.15, according to *The Jakarta Post* (Anon, 2019a): "Roughly half of Jakarta comprises slums, according to a statement from the government." The same article went on to state:

> In January, Vice President Jusuf Kalla and the governor noted the gap between the city's urban and suburban areas. Kalla commented that the main

TABLE 7.15 Land use in three districts in Jakarta

Land use	*Area ha*	*%*	*Area (ha)*	*%*	*Area (ha)*	*%*
	Penjaringan 2016		*Cengkareng 2016*		*Cakung 2015*	
Reservoir	90.69	2.2	3.92	0.1	11.29	0.3
Other public facilities	3.69	0.1	-		48.70	1.2
City forest	34.07	0.8	157.74	5.4	59.05	1.5
Industrial areas	909.39	22.3	446.80	15.2	1094.28	27.8
Settlement	1552.42	38.2	2087.61	70.9	2188.25	55.7
Transportation facilities	90.73	2.2	-		-	
Open zone	993.28	24.5	247.35	8.4	268.34	6.8
River	91.61	2.3	-		-	
Pond	296.43	7.3	-		12.68	0.3
Rice field	-		-		247.12	6.3
Total	4062.31	99.9*	2943.42	100	3929.71	99.9*

* Numbers rounded.

> thoroughfare of Jl. MH Thamrin in Central Jakarta looked like Singapore while areas in Tanjung Priok, North Jakarta, looked like Bangladesh.

As in other parts of the world, slum settlements are usually owner built on marginal land. An earlier report (World Bank 2011:20) noted that a slum in Jakarta was graded based on the following: "the number of houses unfit for habitation, the quality of water and sewer services, population density, the use of streams as toilets and water sources, and how marginal of a location it is found in."

Contrasted against the slums are the high-rise apartment buildings in Jakarta. A 2019 article in *The Jakarta Post* (Anon, 2019b) stated that there were then 209,286 apartments in Jakarta with an occupancy rate of 87%. Savirani and Wilson (2018) note that Jakarta has some 25,000 social housing units, many of which appear to be in walk-up blocks, usually located on land away from the city centre, which causes both problems and loss of income for families moved so the land on which their former homes stood can be developed. Given a 2017 population of 9.63 million and an average household size of 5.06 people (Global Data Lab, 2021) gives approximately 1.9 million households. Assuming half live in slums the remainder are spread between high-rise, walk-up blocks, houses and villas and low-rise kampung housing, which gives the distribution of house types shown in Table 7.16. The evidence for Jakarta being such a low-rise city is shown in Figure 7.1. Additionally, the conversion of land to floor area is a factor of two in Jakarta compared to eight in Singapore (Haryanto, 2019).

In 1990 the floor area per person in Jakarta was 10.2 m^2 (Angel and Mayo, 1995:124). In 2019 modern middle-range houses in Greater Jakarta are reported to be 52–77 m^2 (Cushman and Wakefield, n.d.). However, another slightly earlier report suggested the only houses millennials could afford in Jakarta would be one-bedroom, one-bathroom houses of 35 m^2 (Coconuts Jakarta, 2017). Taken together with the average household size, this suggests an estimate of 10 m^2 per person for those not living in slums would be reasonable. A study of slum housing published in 1996 found "90 per cent of Dhaka' s urban poor households have only one main room while 65 per cent have less than 100 square feet (9.3 square metres)" (Tipple and Ameen, 1999). Given slums are often constructed of ad-hoc materials and in Jakarta are also found under bridges and highways (Grabner-Strobach, 2015), the environmental impact of a slum is the land it sits on, which in the latter case would be zero since it would be included in the impact of infrastructure.

TABLE 7.16 Estimated housing distribution in Jakarta

Dwelling type	*No. of households*	*%*
Slums	950,000	50
Apartments	209,286	11
Walk-up social housing blocks	25,000	1
Low rise	715,714	38

FIGURE 7.1 The low-rise nature of much of Jakarta

A study by Surahman et al. (2015) based on a large sample of houses calculated the average embodied energy for a simple house as 48 GJ, that of a medium house was 178.3 GJ and a luxurious house was 438.8 GJ. The average areas of these houses were 44.2, 107.2 and 213.6 m^2, respectively. This would give an embodied energy of 1.1 GJ/m^2 for simple houses, 1.7 GJ/m^2 for medium houses and 2.1 GJ/m^2 for luxurious houses. Although these figures did not include recycled materials the authors noted that the "informal private sectors play major roles in material recycling activities in Indonesia", and that recycling could reduce the embodied energy by 10–33%. Given a conversion factor of two as noted above (Haryanto, 2019), the footprints of these houses would be approximately 22, 54 and 107 m^2, respectively. Using the values for dwelling types in Table 7.16 and assuming those living in slums only use reclaimed materials, the embodied energy of dwellings in Jakarta is shown in Table 7.17.

TABLE 7.17 Embodied energy of dwelling in Jakarta

Dwelling type	*No. of households*	*EE (GJ/m^2)*	*EE (TJ)*
Slums	950,000	0	0
Apartments	209,286	2.1	91,834
Walk-up social housing blocks	25,000	1.7	4,458
Low rise	715,714	1.1	34,354
Total			130,646 (131PJ)

In a study of slums in Jakarta, the main road was 6 m wide and sub-roads giving access to houses were approximately 2 m wide, although these roads were "only enough for pedestrians and motorcyclists. The road is also used by local residents as a place to interact such as sitting, chatting, playing, besides also being used as a place to park the vehicle and put things (furniture)" (Wijaya et al., 2019). Harmadi et al. (2015) note that roads only account for 7.76% of land in Jakarta "which is well under the ideal conditions for metropolitan cities (e.g. London and Tokyo), where it is supposed to be at 12–15 percent." Table 7.18 gives the 2010 area of roads in Jakarta (Harmadi et al., 2015).

TABLE 7.18 Roading in Jakarta

Road type	*Area (m^2)*	*Road area as % of total area*
Toll	2,472,680	0.37
Primary artery	2,323,110	0.35
Primary collector	739,384	0.11
Secondary artery	8,443,242	1.28
Secondary collector	6,644,660	1.00
Local	30,724,918	4.65
Total	51,347,994	7.76

Using the same value for the embodied energy of roads as for Ahmedabad of 1,768 GJ/km for a 3.5 m wide road gives energy embodied in Jakarta roads as 25.9 PJ.

Greater Jakarta had some 170 km of above-ground railways, mostly double-tracked, at the turn of the century when the population of Greater Jakarta was an estimated 21 million with 8.4 million in Jakarta itself (Hata, n.d.). Proportionally this would give 68 km in Jakarta, which is an underestimate given the fact that railways lead to the central city area. The track width is 1,067 mm (3ft 6in gauge) and the land required for a double track is approximately 15 m (Connor, 2017). This gives an area of approximately 1 million square metres or 102 ha, but as noted in the discussion of Amsterdam, this area can carry many more passengers than the equivalent journey by road. The embodied energy of 68 km of track (8,506 GJ/km) is 578 TJ (0.6 PJ), an underestimate that has to be used given the lack of further data.

In 2019 the new additions to commercial buildings "brought total existing stock in Jakarta CBD up to almost 6.6 million sqm. Of that figure, Grade A remained the largest proportion of the total existing stock (39%), followed by Grade B (28%), Premium Grade (23%), and Grade C buildings (10%)" (Savills World Research, 2020). A case study of 10 office buildings in Hong Kong found initial embodied energy values of 5.36–10.93 GJ/m^2 (Wang et al., 2018). The value of 7.7 GJ/m^2 used for Beijing commercial buildings falls within this range. In the absence of specific data, the value of 7.7 GJ/m^2 was applied to the 6.6 million m^2, to give an embodied energy value of 50,820 TJ (51 PJ).

TABLE 7.19 Estimated embodied energy in the built environment of Jakarta

Category	*EE (PJ)*	*EE/person (GJ)*
Residential	131	12
Industry	218	20
Commercial	51	5
Transport infrastructure	27	2.5
Total	427	39.5

From Table 7.15, in the three areas for which there was land breakdown information, industry accounted for 22.3%, 15.2% and 27.8% of area, making an approximate average of 22%. From Chapter 1, Jakarta had a population of 10.8 million in an area of 662.33 km^2. Applying the 22% would give an area for industry of 132 km^2. Using the same value as previously would give an embodied energy of 217,800 TJ (218 PJ). Given the low values for embodied energy for houses compared with those of Ahmedabad, this is probably an overestimate but in the absence of data it has to stand.

The values for the embodied energy of Jakarta are set out in Table 7.19.

7.5 Comparison and discussion

This analysis has been based on many assumptions and data from different times and is only indicative. However, it still reveals a useful message. Table 7.20 compares the total energy embodied in the built environment, together with the embodied energy per citizen, noting that the figures for Beijing only apply to the area within the fifth ring road.

Obviously if half the population lives in slums, as in the case of Jakarta, the energy embodied in the built environment is going to be low, but with urban development, this energy rises to a level that is not insignificant, but that is also often overlooked when buildings are demolished well before they have reached a state of being uneconomic to maintain and repair. Although the cities were grouped according to the date of their establishment, with the exception of small

TABLE 7.20 Energy embodied in five cities

City	*EE (PJ)*	*PJ/km^2*	*EE/person (GJ)*
Beijing	4893	7.5	489
Amsterdam	319	2.6*	473
Ahmedabad	1374	6.5	247
Singapore	2663	3.8	463.5
Jakarta	327	0.5	39.5

* Area includes water.

heritage areas, this has had little effect on the energy embodied in the buildings and infrastructure of a city. A modern city is a modern city whatever its history. Unfortunately, it is the presence of slums that reduces embodied energy, sending the message that an increase in urbanisation needs to be done once and in such a way that the buildings and infrastructure can be maintained so as to last as long as possible.

References

Abey, S.T. and Anand, K.B. (2019). Embodied Energy Comparison of Prefabricated and Conventional Building Construction. *Journal of the Institution of Engineers (India). Series A, Civil, Architectural, Environmental and Agricultural Engineering 100*(4), 777–790.

African City Planner (n.d.). Nairobi: Mixed Use Zones are Redefining the City. Retrieved from africancityplanner.com, 7 July 2021.

AUDA (Ahmedabad Urban Development Authority) (n.d.). Section C: Planning Regulations. Retrieved from file:///C:/Users/Brenda/Downloads/AUDA%20DP%20-%20Planning%20Regulations.pdf, 7 September 2021.

Amsterdam.info (2021). The History of the Amsterdam Canals. Retrieved from www.amsterdam.info/canals/history/, 4 July 2021.

AmsterdamTips (2021). Amsterdam Metro Map & Guide. Retrieved from www.amsterdamtips.com/amsterdam-metro, 17 September 2021.

Angel, S. and Mayo, S.K. (1995). ASEAN Urban Housing Sector Performance: A Comparative Perspective. In McGee, T.G. and Robinson, I.M. (eds), *The Mega-Urban Regions of Southeast Asia*. Vancouver: UBC Press, 109–132.

Annez, P.C., Bertaud, A., Bertaud, M-A., Bhatt, B., Bhatt, C., Patel, B. and Phatak, V. (2012). Ahmedabad: More but Different Government for "Slum Free" and Livable Cities. World Bank Policy Research Working Paper 6267. Retrieved from https://documents1.worldbank.org/curated/en/526341468267004528/pdf/wps6267.pdf, 27 July 2021.

Anon (2015). More Than Half of Beijing Residents Live Far from City Center. *Wall Street Journal*. Retrieved from www.wsj.com/articles/BL-CJB-26909, 18 August 2021.

Anon (2018). The First Underground *"train ferry"* from Beijing Railway Station to West Railway Station to open the Diameter Line on the 20th. Retrieved from http://news.sohu.com/20150309/n409529535.shtml, 16 August 2021.

Anon (2019a). Slums Remain a Fact of Life in Jakarta, Ministry Finds. *Jakarta Post*. Retrieved from www.thejakartapost.com/news/2019/05/28/half-of-jakarta-is-slum-ministry-says.html, 5 August 2021.

Anon (2019b). Indonesian Property Developers Warned About Risks of Apartment Expansion. *Jakarta Post*. Retrieved from www.thejakartapost.com/news/2019/10/16/indonesian-property-developers-warned-about-risks-of-apartment-expansion.html, 6 August 2021.

Anon (2021). Beijing Suburban Railway. Retrieved from https://en.wikipedia.org/wiki/Beijing_Suburban_Railway, 16 August 2021.

ASEANstats (2018). Total Road Length (in kilometer). Retrieved from https://data.aseanstats.org/indicator/ASE.TRP.ROD.A.001, 24 September 2021.

Atlas of Urban Expansion (2016). Beijing, Beijing. Retrieved from Atlas of Urban Expansion, 30 June 2021.

Beijing Municipal Bureau of Statistics (2020). *Beijing Statistical Yearbook 2020*. Retrieved from Beijing Statistical Yearbook 2020, 26 August 2021.

Bharti, M. and Mehrotra, S. (2020). *Metropolitan Ahmedabad, India Scaling Up with Contiguous Replication of Town Planning Schemes*. Washington, DC: World Bank Group. Retrieved from https://openknowledge.worldbank.org/bitstream/handle/10986/34821/Metropolitan-Ahmedabad-Scaling-Up-with-Contiguous-Replication-of-Town-Planning-Schemes.pdf, 2 August 2021.

Carr, C. (2016). Slum-free Cities? How the Women of Ahmedabad Led a Housing Revolution. Retrieved from www.theguardian.com/cities/2016/apr/14/slum-women-ahmedabad-india-housing-revolution, 6 September 2021.

CBS (2021). Small and Relatively Expensive Housing in Amsterdam. Retrieved from www.cbs.nl/en-gb/news/2016/14/small-and-relatively-expensive-housing-in-amsterdam, 4 July 2021.

Census Population (2021). Ahmedabad City Population 2011–2021. Retrieved from www.census2011.co.in/census/city/314-ahmedabad.html, 7 September 2021.

Chang, C.C., Shi, W., Mehta, P. and Dauwels, J. (2019). Life Cycle Energy Assessment of University Buildings in Tropical Climate. *Journal of Cleaner Production 239*, 1–13.

Chang, X., Li, H., Rong, J., Chen, X. and Wang, Y. (2019). Determining the Appropriate Lane Width at Urban Signalised Intersections – a Case Study in Beijing. *IET Intelligent Transport Systems 13*(12), 1785–1791.

Chang, Y., Ries, R.J. and Lei, S. (2012). The Embodied Energy and Emissions of a High-rise Education Building: A Quantification Using Process-based Hybrid Life Cycle Inventory Model. *Energy and Buildings 55*, 790–798.

Chau, C., Soga, K., O'Riordan, N. and Nicholson, D. (2012). Embodied Energy Evaluation for Sections of the UK Channel Tunnel Rail Link. *Proceedings of the Institution of Civil Engineers—Geotechnical Engineering 165*(2), 65–81.

Chew, R. (2019). Five Things You Probably Did Not Know About the Size of Malaysian Homes. Retrieved from www.edgeprop.sg/property-news/five-things-you-probably-did-not-know-about-size-malaysian-homes?__cf_chl_captcha_tk__=pmd_vNnOD.KYXH29O7VdRnNcERsSOHWidgXhYKiADRW51IE-1632274166-0-gqNtZGzNA1CjcnBszQil, 22 September 2021.

Coconuts Jakarta (2017). Survey finds only 5% of millennials in Jakarta will be able to afford a house by 2020 as property prices continue to soar. Retrieved from https://coconuts.co/jakarta/news/survey-finds-5-millennials-jakarta-will-able-afford-house-2020-property-prices-continue-soar/, 9 August 2021.

Connor, P. (2017). Track Basics. Retrieved from Microsoft Word—Track Basics v2.docx (railway-technical.com), 9 August 2021.

Cushman and Wakefield (n.d.). Marketbeat: Greater Jakarta. Retrieved from file:///C:/Users/Brenda/Downloads/Indonesia--Jakarta--Landed-Residential-Q4-2019.pdf, 6 August 2021.

de Wilde, S. and van den Dobbelsteen, A. (2004). Space Use Optimisation and Sustainability—Environmental Comparison of International Cases. *Journal of Environmental Management 73*(2), 91–101.

Debnath, A., Singh, S.V. and Singh, Y.P. (1995). Comparative Assessment of Energy Requirements for Different Types of Residential Buildings in India, *Energy and Buildings 23*, 141–146.

Demographia (2000). City of New York: Land Use. Retrieved from City of New York: Land Use (demographia.com), 5 July 2021.

Du, J. (2018). Beijing takes aim at congestion, pollution with new car limits. Retrieved from www.chinadaily.com.cn/a/201806/16/WS5b244a9fa310010f8f59d4a0.html, 1 July 2021.

DVV Media International (2019). Beijing Metro now the world's longest. Retrieved from www.railwaygazette.com/projects-and-planning/beijing-metro-now-the-worlds-longest/55436.article, 16 August 2021.

EuroTest (n.d.). Road Works 2007 Guidelines in Europe: When in Rome.....Retrieved from https://web.archive.org/web/20120311004223/http://www.eurotestmobility.net/eurotest.php?itemno=224&lang=EN, 23 August 2021.

Feddes, F., de Lange, M. and te Brömmelstroet, M. (2020). Hard Work in Paradise. The Contested Making of Amsterdam as a Cycling City. In Cox, P. and Koglin, T. (eds), *The Politics of Cycling Infrastructure: Spaces and (In)Equality*. Bristol: Policy Press.

Fleischer, F. (2010). *Suburban Beijing: Housing and Consumption in Contemporary China.* Minneapolis: University of Minnesota Press.

Gao, Y. and Newman, P. (2018). Beijing's Peak Car Transition: Hope for Emerging Cities in the 1.5 °C. *Urban Planning 3*(2), 82–93.

Global Data Lab (2021). GLD Area Profile Report: DKI Jakarta. Retrieved from https://globaldatalab.org/profiles/region/IDNr110/, 6 August 2021.

Government of Singapore (2021). Rail Network. Retrieved from www.lta.gov.sg/content/ltagov/en/getting_around/public_transport/rail_network.html, 24 September 2021.

Grabner-Strobach, A. (2015). Slums. Retrieved from www.humedica.org/articles/2020/jakarta-between-skyscrapers-and-slums/index_eng.html, 9 August 2021.

Graham, E. (2021). Upsizing Singapore's Defence Footprint in Australia. Retrieved from www.lowyinstitute.org/the-interpreter/upsizing-singapores-defence-footprint-australia-0, 26 July 2021.

GVB (2013). About GVB: Network. Retrieved from https://archive.is/20141212181619/http://en.gvb.nl/overgvb/bedrijfsprofiel/MaterieelEnCijfers/Pages/Netwerk.aspx/, 17 September 2021.

Guilford, G. (2014). Parking in some areas of Beijing now costs more than $160,000. Retrieved from https://qz.com/212897/parking-in-some-areas-of-beijing-now-costs-more-than-160000/, 19 August 2021.

Harmadi, S.H.B., Yudhistira, M.H. and Koesrindartono, D.P. (2015). How Does Congestion Matter for Jakarta's Citizens? *Journal of Indonesian Economy and Business 30*(3), 220–239.

Haryanto, W. (2019). Too Dense: An Excuse for Jakarta's Inaction! *Jakarta Post*. Retrieved from www.thejakartapost.com/academia/2019/10/08/too-dense-an-excuse-for-jakartas-inaction.html, 6 August 2021.

Hata, T. (n.d.). Improvement of Railway System in Jakarta Metropolitan Area. Retrieved from www.ejrcf.or.jp/jrtr/jrtr35/f36_hat.html, 9 August 2021.

Herlambang, S., Leitner, H., Ju Tjung, L., Sheppard, E. and Anguelov, D. (2019). Jakarta's Great Land Transformation: Hybrid Neoliberalisation and Informality. *Urban Studies 56*(4), 627–648.

Hickman, R., Fremer, P., Breithaupt, M. and Saxena, S. (2011). *Changing Course in Urban Transport, An Illustrated Guide*. Manila: Asia Development Bank/Eschborn: Deutsche Gesellschaft für Internationale Zusammenarbeit (GIZ) GmbH.

Jacobson, M.Z., Nghiem, S.V., Sorichetta, A. and Whitney, N. (2015). Ring of Impact from the Mega-urbanization of Beijing Between 2000 and 2009. *Journal of Geophysical Research: Atmospheres 120*, 5740–5756.

Karlsson, C.S.J., Miliutenko, S., Björklund, A., Mörtberg, U., Olofsson, B. and Toller, S. (2017). Life Cycle Assessment in Road Infrastructure Planning Using Spatial Geological Data. *The International Journal of Life Cycle Assessment 22*, 1302–1317.

Khan, A.M. (2014). Revisiting Planning Standards for Recreational Facilities in Urban Areas. Retrieved from www.bip.org.bd/admin/uploads/bip-publication/publication-12/paper/20141118151124.pdf, 28 August 2021.

Khan, M.E. and Zheng, S. (2016). *Blue Skies over Beijing: Economic Growth and the Environment in China*. Princeton, NJ: Princeton University Press.

Kodukula, S., Rudolph, F., Jansen, U. and Amon, E. (2018). *Living. Moving. Breathing.* Wuppertal: Wuppertal Institute.

Koezjakov, A., Urge-Vorsatz, D., Crijns-Graus, W. and van den Broek, M. (2018). The Relationship Between Operational Energy Demand and Embodied Energy in Dutch Residential Buildings. *Energy and Buildings 165*, 233–245.

Kua, H.W. and Wong, C.L. (2012). Analysing the Life Cycle Greenhouse Gas Emission and Energy Consumption of a Multi-storied Commercial Building in Singapore from an Extended System Boundary Perspective. *Energy and Buildings 51*, 6–14.

Li, F., Zhang, F., Li, X., Wang, P., Liang, J., Mei, Y., Cheng, W. and Qian, Y. (2107). Spatiotemporal Patterns of the Use of Urban Green Spaces and External Factors Contributing to Their Use in Central Beijing. *International Journal of Environmental Research and Public Health 14*(3), 17pp.

Li, X., Yuan, D., Guo, Y. and Cai, Z. (2016). Use of a 10.22 m diameter EPB shield: a case study in Beijing subway construction. *SpringerPlus 5*:2004. Retrieved from https://springerplus.springeropen.com/track/pdf/10.1186/s40064-016-3672-5.pdf, 20 August 2021.

Li, Y.L., Han, M.Y., Liu, S.Y. and Chen G.Q. (2019). Energy Consumption and Greenhouse Gas Emissions by Buildings: A Multi-scale Perspective. *Building and Environment 151*, 240–250.

Lo, R.H. (2010). The City as a Mirror: Transport, Land Use and Social Change in Jakarta. *Urban Studies 47*(3), 529–555.

Long, Y. and Zhang, E. (2021). *Data Augmented Design: Embracing New Data for Sustainable Urban Planning and Design*. Cham: Springer Nature.

Macrotrends (2021). *New York City Metro Area Population 1950–2021*. Retrieved from www.macrotrends.net/cities/23083/new-york-city/population, Accessed 5 July 2021.

Miller, D., Doh, J.H., Lima, M.M. and van Oers, N. (2014). Embodied Energy Assessment of the Structural System in Concrete Buildings: A Case Study of 7 South-east Queensland Structures. Proceedings 23rd Australasian Conference on the Mechanics of Structures and Materials (ACMSM23) Byron Bay, Australia, 9–12 December. 1217–1222.

Ministry of MSME (Micro, Small and Medium Enterprises) (n.d.). Brief Industrial Profile of Ahmedabad District. Retrieved from http://dcmsme.gov.in/old/dips/dip%20ahmedabad_gu.pdf, 6 September 2021.

Ministry of National development (2013). A High Quality Living Environment for All Singaporeans: Land Use Plan to Support Singapore's Future Population. Retrieved from www.mnd.gov.sg/docs/default-source/mnd-documents/publications-documents/land-use-plan.pdf, 26 July 2021.

Ministry of Urban Development (n.d.). City Profile: Ahmedabad. Retrieved from http://utbenchmark.in/UsersidePages/CityProfile.aspx?City=1, 2 August 2021.

Mithraratne, M. (2011). Lifetime Liabilities of Land Transport Using Road and Rail Infrastructure. NZ Transport Agency research report 462.

Nairobi Slums project (2021). *Nairobi Slums*. Retrieved from Nairobi Slums, 7 July 2021.

NationMaster (2021). Transport > Road density > Km of road per 100 sq. km of land area: Countries Compared. Retrieved from www.nationmaster.com/country-info/stats/Transport/Road-density/Km-of-road-per-100-sq.-km-of-land-area, 23 August 2021.

Netherlands Insiders (2021). Why Are There So Many Canals in Amsterdam. Retrieved from Netherlands Insiders, 4 July 2021.

Nie, W., Greeven, M.J., Feng, Y. and Wang, J. (2022). *The Future of Global Retail: Learning from China's Retail Revolution*. Abingdon: Routledge.

NUS (2019). University Campus Infrastructure. Retrieved from https://uci.nus.edu.sg/wp-content/uploads/2019/12/FA_190419_INOUT2Aug2019.pdf, 22 September 2021.

OECD (2017). *Land-use Trends, Challenges and Opportunities in Amsterdam*. Paris: OECD Publishing.

Onderzoek, Informatie en Statistiek (2019). *Gemeente Amsterdam: Stadsdelen in cijfers 2019*. Retrieved from https://data.amsterdam.nl/publicaties/publicatie/stadsdelen-in-cijfers-2019/f4b55c38-6b3c-4143-bea5-94c4f1cb0887/, 17 September 2021.

Oyugi, M.O. and K'Akumu, O.A. (2007). Land Use Challenges for the City of Nairobi. *Urban Forum 18* (1), 94–113.

Pan, L., Zhu, M., Lang, N. and Huo, T. (2020). What Is the Amount of China's Building Floor Space from 1996 to 2014? *International Journal of Environmental Research and Public Health 17*(16), 5967.

Patel, S. (2016). Public Spaces for All: How "Public" Are Public Spaces? Dissertation for Bachelor in Planning, CEPT University, Ahmedabad.

Pavement Interactive (n.d.). Energy and Road Construction—What's the Mileage of Roadway? Retrieved from https://pavementinteractive.org/energy-and-road-construction-whats-the-mileage-of-roadway/, 23 August 2021.

Peng, G.T.C. (1972). The Philosophy of the City Design of Peking. *Ekistics 33*(195), 124–129.

Penadés-Plà, V., García-Segura, T. and Yepes, V. (2019). Accelerated Optimization Method for Low-embodied Energy Concrete Box-girder Bridge Design. *Engineering Structures 179*, 556–565.

Putri, R.F., Wibirama, S., Giyarsih, S.R., Pradana, A. and Kusmiati, Y. (2019). Landuse Change Monitoring and Population Density Analysis of Penjaringan, Cengkareng, and Cakung Urban Area in Jakarta Province. The 4th International Conference on Science and Technology (ICST 2018). Retrieved from www.e3s-conferences.org/articles/e3sconf/abs/2019/02/e3sconf_icst2018_03004/e3sconf_icst2018_03004.html, 4 August 2021.

Rodrigues, V., Martins, A.A., Nunes, M.I., Quintas, A., Mata, T.M. and Caetano, N.S. (2018). LCA of Constructing an Industrial Building: Focus on Embodied Energy and Carbon. *Energy Procedia 153*, 420–425.

Russo, A. and Cirella, G.T. (2018). Modern Compact Cities: How Much Greenery Do We Need? *International Journal Environmental Research and Public Health 15*(10), 2180.

Savills World Research (2020). Spotlight Office: Jakarta. Retrieved from https://pdf.savills.asia/asia-pacific-research/indonesia-research/jakarta-office-feb-2020.pdf, 10 September 2021.

Savirani, A. and Wilson, I. (2018). Distance Matters: Social Housing for the Poor. Retrieved from www.insideindonesia.org/distance-matters-social-housing-for-the-poor, 6 August 2021.

Sharma, R. (2021). Explained: The Sabarmati River Front Development Project, and What it Seeks to Achieve. Retrieved from https://indianexpress.com/article/explained/explained-what-is-the-sabarmati-river-front-development-project-what-it-seeks-to-achieve-7249497/, 2 September 2021.

Singapore Government Statistics (n.d). Road length In Kilometer (end of year). Retrieved from www.lta.gov.sg/content/dam/ltagov/who_we_are/statistics_and_publications/statistics/pdf/Road-Length-km.pdf, 24 September 2021.

Singh, A. Vaddy, P. and Biligiri, K.P. (2020). Quantification of Embodied Energy and Carbon Footprint of Pervious Concrete Pavements Through a Methodical Lifecycle Assessment Framework. *Resources, Conservation and Recycling 161*, 13pp.

Singh, M.K. (2010). UPA-II to Unveil "Slum-Free India" Dream Scheme. *The Times of India*. 18 October 2010.

SingHealth Group (2021). SGH Campus Masterplan. Retrieved from www.sgh.com.sg/about-us/corporate-profile/Pages/SGH-Campus-Masterplan.aspx, 24 September 2021.
Slotta, D. (2021). Per Capita Living Space in Urban Beijing 2008–2019. Retrieved from www.statista.com/statistics/1135698/china-per-capita-living-space-in-urban-areas-of-beijing/, 26 August 2021.
Song, C. (2021). Beijing Subway. Retrieved from www.chinahighlights.com/beijing/transport/beijing-subway/, 16 August 2021.
Statista Research Department (2021). Total Supply of Office Space with More Than 500 Square Meters of Lettable Floor Area in Amsterdam in the Netherlands from 2012 to 2020. Retrieved from www.statista.com/statistics/629259/supply-square-meters-of-lettable-floor-area-amsterdam-in-the-netherlands/, 17 September 2021.
Studio Meem (2015). Mapping Cairo. Retrieved from Cairobserver—Mapping Cairo, 3 August 2021.
Su, X. and Zhang, X. (2016). A Detailed Analysis of the Embodied Energy and Carbon Emissions of Steel-construction Residential Buildings in China. *Energy and Buildings 119*, 323–330.
Surahman, U., Kubota, T. and Higashi, O. (2015). Life Cycle Assessment of Energy and CO_2 Emissions for Residential Buildings in Jakarta and Bandung, Indonesia. *Building 5*(4), 1131–1155.
Tipple, G. and Ameen, M.S. (1999). User Initiated Extension Activity in Bangladesh: "Building Slums" or Area Improvement? *Environment and Urbanization 11*(1), 165–183.
Travel China Guide (2021). Beijing Hutong. Retrieved from www.travelchinaguide.com/cityguides/beijing/hutong/, 28 June 2021.
Udajea, C., Trillo, C., Awuah, Kwasi G.B., Makore, Busisiwe, C.N., Patel, D.A., Mansuri, Lukman, E. and Jha, Kumar, N. (2020). Urban Heritage Conservation and Rapid Urbanization: Insights from Surat, India. *Sustainability 12*(6), 26pp.
UNHabitat (2013). Street as Public Spaces and Drivers of Urban Prosperity. Retrieved from https://unhabitat.org/sites/default/files/2020/08/streets_as_public_spaces_and_drivers_of_urban_prosperity.pdf, 19 August 2021.
Van Lohuizen, H.P.S. (1989). A new pre-metro line for Amsterdam? *Tunnelling and Underground Space Technology 4*(3), 285–291.
Venkatarama Reddy, B.V. and Jagadish, K. S (2003). Embodied Energy of Common and Alternative Building Materials and Technologies. *Energy and Buildings 32*(2), 129–137.
Verdict Media Ltd. (2021). Ahmedabad Metro Rail, Gujarat, India. Retrieved from www.railway-technology.com/projects/ahmedabad-metro-rail-gujarat/, 7 September 2021.
Wang, H., Qiu, J., Breuste, J., Friedman, C R., Zhou, W. and Wang, X. (2013). Variations of urban greenness across urban structural units in Beijing, China. *Urban Forestry and Urban Greening 12*. 554–561.
Wang, J., Yu, C. and Pan, W. (2018). Life Cycle Energy of High-rise Office Buildings in Hong Kong. *Energy and Buildings 167*, 152–164.
Wei, K.H. (n.d.). Urban Metabolism (Industrial Ecology) Studies. Retrieved from www.ura.gov.sg/uol/-/media/User%20Defined/URA%20Online/urban-lab/explore-research/themes/environmental-quality/urban-metabolism-executive-summary.pdf?la=en, 24 September 2021.
Whitehand, J.W.R. (1994). Development Cycles and Urban Landscapes. *Geography 79*(1), 3–17.
Whitelegg, J. (1994). Transport and Land Take: A Report for CPRE. Retrieved from www.eco-logica.co.uk/pdf/CPRELandTake.pdf, 1 July 2021.

WHO (2016). *Urban Green Spaces and Health*. Copenhagen: WHO Regional Office for Europe.

Wijaya, A., Ardalia, F. and Dewi, E.P. (2019). Pemanfaatan Ruang Komunal Pada Kawasan Permukiman Kumuh Perkotaan Di Manggarai Jakarta Selatan. *Jurnal Sains & Teknologi 3*(2), 17–26.

World Bank (2011). Jakarta—Urban Challenges in a Changing Climate (English). Washington, DC: World Bank Group. Retrieved from http://documents.worldbank.org/curated/en/132781468039870805/Jakarta-Urban-challenges-in-a-changing-climate, 5 August 2021.

Wu, Z., Yao, L., Zhuang, M. and Ren, Y. (2020). Detecting Factors Controlling Spatial Patterns in Urban Land Surface Temperatures: A Case Study of Beijing. *Sustainable Cities and Society 63*, 9pp.

Xinhua (2017). *China's per capita floor area exceeds 40 square meters*. Retrieved from China's per capita floor area exceeds 40 square meters - Business - Chinadaily.com.cn, Accessed 28 June 2021.

Xinhua (2019). *Total length of roads in Beijing surges 130 times in past 70 years*. Retrieved from Total length of roads in Beijing surges 130 times in past 70 years - Xinhua | English. news.cn (xinhuanet.com), Accessed 30 June 2021.

Yang, J., Huang, C., Zhang, Z. and Wang, L. (2014). The temporal trend of urban green coverage in major Chinese cities between 1990 and 2010. *Urban Forestry & Urban Greening 13*(1). 19–27.

Zhang, H.H. and Brown, D.F. (2005). Understanding Urban Residential Water Use in Beijing and Tianjin, China. *Habitat International 29*(3), 469–491.

Zhang, Y. (2008). Steering Towards Growth: Symbolic Urban Preservation in Beijing, 1990–2005. *Town Planning Review 79*(2–3), 187–208.

Zhao, J., Ouyang, Z., Zheng, H., Zhou, W., Wang, X., Xu, W. and Ni, Y. (2010). Plant Species Composition in Green Spaces Within the Built-up of Beijing, China. *Plant Ecology 209*(2), 189–204.

8

URBAN POLICIES

8.1 Introduction

Originally there was no plan to have a chapter about the policies of the cities included in this investigation. The intent was to calculate environmental impact using the physical data of the cities and discuss the results. It meant finding official data for each city, understanding what was and was not included in that data, comparing the information and calculating the environmental impact for each city. Early in the research, the issue with the data was clear, as has also become apparent in the preceding chapters, with problems like no clear boundaries for the city, data collected in different ways and often data simply missing.

During this search for data, it became clear that the cities had policies about their future survival. As a result, a new layer of data was collected—urban policies related to environmental impact. A simple glance at urban policies reveals that many are related to environmental impact and most of the cities in this book have policies directly related to climate change mitigation. Cases like "Pathways to Carbon Neutral New York City" (New York City Council, 2021) or Amsterdam's "Roadmap for a Climate Neutral Amsterdam 2050" (City of Amsterdam, 2021) are very common. Amsterdam for example states what might be perceived as very ambitious: "By 2050, we want to have cut CO_2 emissions by 95% compared to 1990 levels…we want to cut CO_2 emissions by 55% in 2030." (City of Amsterdam, 2019:4). Statements like this are common and may be perceived as hopeful but possibly reachable. In fact, Amsterdam scores very high in our comparative assessment, although not as high as Copenhagen (see Table 8.12). It thus made sense to view the policies side by side and to try to rank them. Considering that since 2009 a majority of the world's population has been living in urban areas (United Nations, 2021a), looking at how cities are trying to reduce their environmental impact and prepare for the future seems to be a good way to compare them.

DOI: 10.4324/9781003045847-9

8.2 Methodology

To produce a meaningful comparison, an analytical tool had to be developed. This was done by dividing the data on city policies into four groups, each with a number of subgroups, and points were awarded following a predetermined standard (Table 8.1). As an example, in the Policies group, the subgroups are Clarity of priorities, Measurability of goals and Feasibility of achievement (Table 8.2). For this group with its subgroups, points were then awarded under the two headings of Transportation and Energy (Table 8.1). The same two headings appear in the Collaboration group (Table 8.1). This looks like double counting, but the decision to do this was made because of the environmental weight of the items and because a city may have very good policies for CO_2 emissions in transportation but nothing regarding the domestic energy mix. The data suggested that city authorities identified energy and transportation as the main contributors to their CO_2 emissions, so energy and transportation were assessed separately under both Policies and Collaboration.

Not all the groups have the same importance when it comes to reducing environmental impact (Table 8.1).

TABLE 8.1 Summary of criteria for assessment of policies

Criteria	• Policies—3 subgroups from 1 to 5 points (energy and transportation); Possible total: 30 points • Collaboration—2 subgroups from 1 to 3 points (energy and transportation); Possible total: 12 points • Government Structure—2 subgroups from 1 to 3 points; Possible total: 6 points • Political Will—3 subgroups: 1 group from 1 to 5 points and 2 subgroups from 1 to 3 points; Possible total: 11 points

8.2.1 Policies

The Policies group has a potential total of 30 points. The three subgroups were evaluated and points given from 1 to 5. Up to 15 points were given to the three subgroups as they relate to Energy, and similarly up to 15 points were given to Transportation as shown in Table 8.2.

TABLE 8.2 Subheadings under Policies

Policies	*Energy points*	*Transportation points*
1. Clarity of priorities	1 to 5	1 to 5
2. Measurability of goals	1 to 5	1 to 5
3. Feasibility of achievement	1 to 5	1 to 5
Total score	Up to 15	Up to 15

This first group refers to the efficiency of the policy, not necessarily its current environmental achievements. It considers how clear, understandable, and applicable the policy is. The criteria were based on Volkery et al. (2004) and Lafferty et al. (2007). Volkery et al. (2004) refer to the clarity of the priorities, and if the goals are well explained and defined, while Lafferty et al. (2007) confirm the importance of realistic and measurable goals.

The first subgroup, Clarity of priorities, assesses how clear the priorities are in terms of allowing the policy to be understood and engaged with by a wide audience, in this case, the citizens, companies and public institutions of each city. The criteria for scoring are set out in Table 8.3.

The second subgroup refers to how measurable the parameters of the policy are (Table 8.4). Measurements might include the share of renewable energy sources, size of electric vehicles fleet or direct CO_2 emissions.

The third subgroup refers to how realistic the goals and subgoals are and the real capacity to achieve them (Table 8.5). It also assesses how the cities have performed in achieving previous targets. This item was created because of the way the cities present their goals. For instance, Copenhagen aims to become "carbon-neutral" by 2025. To achieve this, there are several subgoals, such as a 20% cut in heat consumption; a 20% cut in power consumption in businesses; a 10% reduction in power consumption in households compared to 2010; and installation of solar systems to generate 1% of electricity (European Union, 2015). Another example is Dubai, which aims to have an energy mix comprising solar energy (25%), nuclear power (7%), clean coal (7%) and gas (61%) by 2030 (UEA, 2021). However, what is more important is not the intention but achievements against past goals in these areas.

TABLE 8.3 Policy clarity criteria

Points	*Are clear priorities set?*
1	The strategy does not align with any clear priorities.
2	Only some priorities within the strategy are made clear.
3	Most of the priorities within the strategy are clear.
4	The strategy aligns with clear priorities.
5	The strategy aligns with clear priorities, and also with other indicators such as time constraints.

TABLE 8.4 Measurability of policy priorities

Points	*Are goals measurable?*
1	Goals do not align with any measurable indicators.
2	Goals include a few measurable indicators.
3	Goals are equally based on measurable and not measurable indicators.
4	Goals are mostly comprised of measurable indicators.
5	Goals are very specific and easily measurable.

TABLE 8.5 Criteria for feasibility of achieving goals

Points	*Are goals realistic?*
1	Subgoals are likely not to be met and goals in the past have not been met.
2	Subgoals are likely not to be met.
3	Subgoals are likely to be met.
4	Linear growth has been considered making meeting subgoals likely.
5	Goals in the past have been met, linear growth has been considered and subgoals are likely to be met.

8.2.3 Collaboration

The Collaboration group has a potential total of 12 points. The two subgroups were assessed and points are given from 1 to 3. Up to 6 points were given to the two subgroups as they relate to Energy and up to 6 points were similarly given to Transportation as shown in Table 8.6.

TABLE 8.6 Subheadings under Collaboration

Collaboration	*Energy points*	*Transportation points*
1. Citizen participation	1 to 3	1 to 3
2. Government Collaboration	1 to 3	1 to 3
Total score	Up to 6 points	Up to 6 points

The second and third groups of criteria, Collaboration and Government Structure, were both divided into two subgroups with up to three points for each criterion. The Collaboration group was based on Krause et al. (2016), who discussed the importance of public collaboration, focussing on citizen participation and collaboration between governments and external players. As a result, the Collaboration group evaluated citizen participation and the interaction with external partners such as universities, businesses and non-governmental organisations (NGOs) (Tables 8.7 and 8.8).

TABLE 8.7 Citizen participation criteria

Points	*Is citizen participation part of the strategy?*
1	Citizen participation is not part of the strategy.
2	Citizen participation is part of the strategy to some extent.
3	Citizen participation is an important aspect of the strategy.

TABLE 8.8 Government collaboration criteria

Points	Does the government collaborate with outside actors other than citizens? (e.g. businesses, universities, NGOs)
1	The government does not collaborate with other outside actors.
2	The government occasionally collaborates with other outside actors.
3	The government has strong connections with other outside actors.

8.2.4 Government Structure

The remaining groups do not have the complication of having the two headings of Energy and Transportation and the points are awarded directly (Table 8.9).

TABLE 8.9 Subheadings under Government Structure

Government Structure	*Points*
1. Active participation of city in network(s)	1 to 3
2. Governmental body for sustainable development	1 to 3
Total score	Up to 6 points

The Government Structure group addresses the issue of how the city government appraises and create its environmental policies. It is based on the work of Dalal-Clayton and Bass (2002, 2006), European Commission (2004) and Volkery et al. (2004) and includes the relationship between the cities in networks, as explained by Evans et al. (2005) and Kern and Gotlind (2009). Such networks are based on best-practice transfer and learning strategies among their members. Tables 8.10 and 8.11 set out the criteria for awarding points.

TABLE 8.10 Network criteria

Points	*Is the city part of a network?*
1	The city is not part of any network.
2	The city is part of at least one network but this is not active.
3	The city is an active member of at least one network.

TABLE 8.11 Government body for sustainable development criteria

Points	*Is sustainable development incorporated within a regulatory body or does it form its own? (Independent body for environmental issues)*
1	Sustainable development is not incorporated within any specific governmental body but is considered from time to time by different unspecified bodies.
2	Sustainable development is incorporated into specified existing governmental bodies.
3	Sustainable development forms a separate body of its own.

8.2.5 Political will

The last group of assessments refers to Political Will. This was mentioned by Kern and Gotlind (2009) and Evan et al. (2005) as an important factor in achieving meaningful change. The same sources also pointed to the relevance of the display of willingness to promote environmental changes. Because of this, five points could be awarded based on how high sustainability is on the political agenda, and the influence of lobbying on the development of environmental policies. Sühlsen and Hisschemöller (2014) support the idea of including lobbying as they consider this a relevant factor for urban policies.

Tables 8.13 to 8.15 provide details of the subgroups within the Political Will group.

The data for every city was collected in the same way using official sources such as websites and reports wherever possible. In some cases, when the information was not easily found in official channels, sources such as reliable news outlets were used. Data that did not come from official sources were analysed and the sources verified before a decision was made about using it. The data was collected by six researchers from three different countries.[1]

Although this part of the research used a mixed-methods approach, qualitative analysis was primarily applied. Quantitative data was not extensively available, and most of the policies collected for this research did not present their calculation methodology along with the relevant policies. The assessment of the policies is, therefore, a qualitative analysis that occasionally uses quantitative data, such as production of renewable energy, or reduction in fossil fuel use. Even so, some quantitative data may rely on too many assumptions to be taken as a numerical indicator.

TABLE 8.12 Subheadings and points under Political Will

Political Will	*Points*
1. Entrepreneurial mayor, green party etc.	1 to 5
2. Number of agendas	1 to 3
3. Connection of politicians to lobbyists, influence of lobbyism in the past	1 to 3
Total score	Up to 11

TABLE 8.13 Political willingness criteria

Points	*Do political staff display willingness for sustainable development?*
1	Political staff do not display willingness for sustainable development.
2	Only a few members display willingness for sustainable development.
3	The political staff is divided into two equal groups of those who display willingness and those who do not.
4	Most of the political stuff display a willingness for sustainable development.
5	Political staff are enthusiastic and display firm willingness for sustainable development.

TABLE 8.14 Importance of sustainability in the political agenda criteria

Points	*How important is sustainability within the political agenda?*
1	Sustainability is not an important topic on the political agenda.
2	Sustainability is taken into account from time to time.
3	Sustainability is an important topic on the political agenda, being frequently taken into account.

TABLE 8.15 Lobbyism criteria

Points	*Does lobbyism influence the political agenda?*
1	Lobbyism has a strong influence on the political agenda.
2	Lobbyism has a small influence on the political agenda.
3	Lobbyism does not influence the political agenda.

The total points score available for the whole Policies Index is
Policies: 30
Collaboration: 12
Government structure: 6
Political will: 11
Total: 59 points.

An example would be the reduction of CO_2 emissions in urban areas, which is normally estimated rather than being obtained through collection of physical data. Table 8.16 sets out a detailed example for scoring Buenos Aires under the Policies group. The total score for Buenos Aires under Policies is 22 points. Table 8.17 shows the scoring under all groups and subgroups for Copenhagen, where the total score is 53 points out of a possible 59.

TABLE 8.16 Scoring Buenos Aires for Policies

Policies	*Energy score*	*Transportation score*	*Explanation for score under Energy*	*Explanation for score under Transportation*
1. Clarity of priorities	3	3	Argentina has a clear renewable energy development strategy, which clearly lists the priorities, goals, and completion time [1,2]	Buenos Aires lacks a detailed development strategy for the future. Since the sustainable transportation plan in 2013, there is no other complete development strategy [3,4]

(continued)

TABLE 8.16 Cont.

Policies	*Energy score*	*Transportation score*	*Explanation for score under Energy*	*Explanation for score under Transportation*
2. Measurability of goals	4	4	Argentina's energy strategy uses greenhouse gas emissions and proportion of renewable energy as measurement indicators [1,2]	There are some indicators for reduction in emissions from transportation [3,4]
3. Feasibility of achievement	4	4	Buenos Aires had a target of 20% renewable electricity for the city's largest public sector users by 2025, which it has already surpassed with renewables making up 32% of the electricity mix in 2019 [1,2]	Buenos Aires implemented several impressive sustainable transport projects in 2013. For its success in promoting urban mobility, reducing emissions, and improving safety, Buenos Aires received the 2014 Sustainable Transport Award. There is a lack of follow-up strategic plans. New indicators such as use of biofuels lack a strong plan to sustain the development, although theoretically it is achievable [3,4]
Total	11	11		

1 Nascimento et al. (2020).
2 Kurdziel et al. (2020).
3 Chitturi (2021).
4 Buenos Aires Ciudad (2022).

TABLE 8.17 Points scored for Policies Index (PI) for Copenhagen

Group	*Subgroup*	*How points are earned*	*Energy*	*Transportation*
Policies	Clarity of priority	Are clear priorities set?	5	5
	Measurability	Are goals measurable?	4	4
	Feasibility	Are goals realistic?	3	4
Collaboration	Citizen participation	Is citizen participation part of the strategy?	3	3
	Government collaboration	Does the government collaborate with outside actors other than citizens?	3	3
Government Structure	Network criteria	Is the city part of a network?	3	
	Government regulatory body	Is sustainable development incorporated within a regulatory body or does it form its own?	3	
Political Will	Political willingness towards sustainability	Do political staff display willingness for sustainable development?	5	
	Sustainability and Political agenda	How important is sustainability within the political agenda?	3	
	Lobbyism	Does lobbyism clearly and openly influence the political agenda?	2	

8.3 Results

It was not possible to collect sufficient data from Havana and Pyongyang to provide a final index for these cities. Overall, most of the cities achieved a high score (Table 8.18), with Copenhagen having the highest and Nairobi the lowest score, noting that Havana and Pyongyang could not be assessed.

From the results above, a series of comparisons were made trying to find correlations between the Policies Index and the Human Development Index (HDI), Democracy Index (DI), Ecological Footprint (EF), CO_2 emissions, and income per capita.

HDI was created as a criterion for assessing development of a country. It includes life expectancy, education and income. It is calculated by the United Nations (United Nations, 2021b). The Democracy Index is based on five categories: electoral process and pluralism, the functioning of government, political participation, political culture and civil liberties and was created by The Economist Intelligence Unit (EIU), which is the research and analysis division of The Economist Group. The Democracy Index uses a criterion created by an international publication, which tends to favour deregulation and privatisation and where the public agenda is liberal in the classical sense (*The Economist*, 2021). In the DI points are awarded according to the criteria above and the countries are classified as full democracies, flawed democracies, hybrid regimes and authoritarian regimes. According to the Economist Intelligence Group (2021:7), the index is deliberately designed to intersect politics, policy, and the economy. This categorisation is important for this book as the ultimate aim is to discuss the capacity democracies have in fighting against poor outcomes for the natural environment. Additionally, this investigation has tried

TABLE 8.18 Policies Index comparison

Cities	*Policies Index score out of a possible 59 points*
Copenhagen	53
Singapore	49.5
Moscow	49
New York	48
Auckland	47
Hamburg	47
Buenos Aires	46
Amsterdam	45
Sydney	45
Beijing	45
Sao Paulo	41
Cairo	40
Jakarta	38
Dubai	36
Kuala Lumpur	36
Ahmedabad	36
Nairobi	31

to shed light on the roles of economic players, who often support privatisation and deregulation, in finding solutions to current environmental issues. From the government point of view, solutions are often based on regulations. It is important to highlight that the aim in using the DI is not to classify the cities according to the political regime. The idea is to use it as a standardised parameter for further analysis. EF was based on material from the Global Footprint Network using national data (Global Footprint Network, 2021a). Finally, the inclusion of CO_2 emissions and national income seemed to be a natural step in assessing the importance of the economy in shaping the policies of the cities.

8.3.1 Correlations

A series of correlations were established using the data in Table 8.19. Correlation, or dependency, evaluates the statistical relationship between two

TABLE 8.19 Comparison of Policies Index (PI), Democracy Index (DI, 2020), Human Development Index (HDI, 2021), Ecological Footprint (EF, 2021), CO_2 emissions (2019) and national average income (2019)

Cities	*Policies Index (out of 59 points)*	*DI*[1]	*HDI*[2]	*EF (gha)*[3]	*CO_2 Emissions/ per capita (tonnes/year)*[4]	*National income (U$$ per capita/ year)*[5]
Copenhagen	53	9.15	9.29	6.9	10.1	59,766.00
Singapore	50	6.03	9.32	5.9	30.8	57,232.00
Moscow	49	3.31	9.31	5.5	6.9	11,262.00
New York	48	7.92	9.24	8	17.1	62,479.00
Auckland	47	9.25	9.17	4.3	6.8	44,678.00
Hamburg	47	8.67	9.36	7.1	11.1	44,791.00
Buenos Aires	46	6.95	8.82	3.3	4.1	13,698.00
Amsterdam	45	8.96	9.31	5.8	10.5	50,921.00
Sydney	45	8.96	9.39	7.3	11.7	57,591.00
Beijing	45	2.27	9.04	3.7	4.2	9,689.00
Sao Paulo	41	6.92	7.59	2.8	1.7	9,991.00
Cairo	40	2.93	7.51	1.9	2.5	2,619.00
Jakarta	38	6.3	6.94	1.7	2.3	4,144.00
Dubai	36	2.7	8.63	9.1	22.3	39,832.00
Kuala Lumpur	36	7.19	8.02	3.9	8.5	10,909.00
Ahmedabad	36	6.61	8.01	1.2	1.8	2,144.40
Nairobi	31	5.1	5.9	1	2.68	1,680.00
Havana	NA	2.84	7.83	1.8	3.60	8,064.00
Pyongyang	NA	1.08	NA	0.8	0.86*	684.00

1 The Economist Intelligence Group (2021).
2 United Nations (2021b).
3 Global Footprint Network (2021a).
4 Moran, et al. (2021); Global Gridded Model of Carbon Footprints (2021).
5 Our World in Data (2021).

TABLE 8.20 Rankings for PI, DI, HDI, income per capita and EF and CO_2 emissions

Cities	*PI rank 1 is highest*	*DI rank 1 is highest*	*HDI rank 1 is highest*	*EF rank 1 is lowest*	*CO_2 rank 1 is lowest*	*Income rank 1 is highest*
Copenhagen	1	2	6	15	13	2
Singapore	2	12	3	14	19	4
Moscow	3	14	5	12	11	10
New York	4	6	7	18	17	1
Auckland	6	1	8	11	10	7
Hamburg	6	5	2	16	15	6
Buenos Aires	7	8	10	8	8	9
Amsterdam	9	4	5	13	14	5
Sydney	9	4	1	17	16	3
Beijing	9	18	9	9	9	13
Sao Paulo	11	9	15	7	2	12
Cairo	12	15	16	6	5	16
Jakarta	13	11	17	4	4	15
Dubai	15	17	11	19	18	8
Kuala Lumpur	15	7	12	10	12	11
Ahmedabad	15	10	13	3	3	17
Nairobi	17	13	18	2	6	18
Havana	19	16	14	5	7	14
Pyongyang	19	19	19	1	1	19

TABLE 8.21 r values and relationships between two sets of data

$0.8 \le \mid r \mid \le 1.0$	Very strong relationship
$0.4 \le \mid r \mid < 0.6$	Moderate relationship
$0.2 \le \mid r \mid < 0.4$	Weak relationship
$0.0 \le \mid r \mid < 0.2$	Very weak relationship

sets of data, demonstrating the strength of a relationship between the two sets (see table 8.21). The correlation coefficient values range between −1.0 and 1.0. Spearman's correlation was selected for this analysis due to the ordinal nature of the data. To apply the Spearman method, a Table 8.20 was created presenting the ranking for each column above. Some assumptions were made to create the rankings and correlations. As the data sets need the same sample size, some assumptions were included. Havana and Pyongyang were given a score of 30 for DI and Pyongyang was assumed to have a HDI of 5.8 just below that of Nairobi. The ranking table was created using descent order for PI, DI, HDI and income per capita, while EF and CO_2 emissions were ranked from the lowest to the highest impact. This means higher EF and CO_2 emissions would rank the cities lower. The idea was to test if a higher PI, DI and HDI would mean lower environmental impact.

From the rankings above, the Spearman rank correlation (Spearman's Rho) was calculated using the following formula:

$$r = 1 - \frac{6\sum d_i^2}{n(n^2 - 1)}$$

where

r = correlation coefficient

d_i = difference between the two ranks of each observation

n = number of observations

8.3.2 Correlation between PI and DI

A perfect positive correlation between PI and DI would give an r value of 1, meaning that if a city ranks top for PI, it would also top the ranking for DI and all other cities would be in the same position in both rankings. A perfect negative correlation ($r = -1$), means that if a city is top for one ranking, it is at the bottom for the other one. If the r value is zero, it means there is no relationship between PI and DI. According to Campbell (2006), the r value determines how strong the relationship is as shown in Table 8.15.

In this case the correlation for PI × DI is $r = 0.54$ (moderate relationship).

The moderate relationship between Policies Index and Democracy Index supports the idea that democracy does not necessarily relate to cities having better policies for reducing environmental impact. This moderate relationship triggers questions that will be explored in Part 2 of this book, such as why democracies and policies do not have a stronger correlation and the factors that affect this relationship.

8.3.3 Correlation of PI and HDI

The r value for PI × HDI of 0.78 indicated a moderate relationship, which is stronger than that between PI and DI. The data suggests that social development, as measured by life expectancy, income and education, is more important in shaping the cities' environmental policies than democracy. However, this relationship is still not a strong one.

8.3.4 Correlation between PI and EF

The r value for the correlation PI × EF was –0.64, indicating a slightly higher but still moderate negative relationship between effective urban policies and environmental impact measured using EF. This means that generally having stronger policies relates only to moderate to low reductions in environmental impacts. An optimistic interpretation of this data could suggest that cities with higher environmental impacts are only recently applying measures to reduce their environmental impact, and therefore, this is not yet reflected in the EF. However, it seems more likely that city policies alone are not capable of tackling the most important aspects of environmental impact and considerably harder and more effective measures are needed to reduce environmental impact.

8.3.5 Correlation between PI and CO_2

Here the r value of –0.54 indicated a moderate negative relationship. As CO_2 emissions form the largest single contributor to EF, with the share increasing every year (Global Footprint Network, 2021b), this should be similar to the relationship between PI and EF.

8.3.6 Correlation between PI and income per capita

Here the r value of 0.78 again indicated an increased but still moderate relationship between PI and income per capita. This was expected, being similar to PI and HDI, as the correlation between HDI and income per capita is strong ($r = 0.85$).

8.4 Discussion

The correlations between PI and DI, HDI, CO_2 emissions, EF and income per capita have shown no strong relationships for the cities included in this investigation. By contrast, as we have seen in earlier chapters, stronger relationships exist for correlations between EF, CO_2 emissions, HDI and income. Higher incomes lead to higher EFs and CO_2 emissions and high HDI indicates higher per capita average incomes.

Although some cities have scored very high in the PI, it is important to highlight what this index is assessing. It was never the goal to check if the cities are achieving what they are promising. However, some criticism can be made of the rationale behind some, or indeed most, of the city policies. First, some of the claims related to CO_2 emissions are very hard to track, calculate and, therefore, verify. This happens primarily because it is technically hard to measure CO_2 emissions. There are too many variables, and often the emissions are simply calculated based on assumptions. It is still not possible to measure emissions at a city scale physically. Also, there are limitations in the science of identifying the origins of CO_2 emissions. It is possible to identify if the source is from burning forests or dying vegetation or if it comes from burning of fossil fuel. This is because CO_2 from burning forests or vegetation contains carbon-14, a radioactive isotope that is generated by cosmic rays hitting the upper atmosphere and then being absorbed into plants, while CO_2 from fossil fuels has virtually no carbon-14 (Pearce, 2021). However, it is still very hard to identify the location of emissions. This brings up another issue related with the origin of emissions, which is recurrent throughout this book, this being how to allocate the emissions for a certain product or service. Should we allocate emissions to the final consumer or the provider or producer of the service? This conundrum helps to explain the claims of carbon neutral cities. It may mean that cities and their citizens are not responsible for the direct release of CO_2 within a city's boundaries, but it does not mean that they are not responsible for emissions. Bakshi (2017) has pointed out the CO_2 impacts for consumption in what he called "hidden emissions" referring to the impact associated with embodied energy, particularly in imported products. This has been discussed in Chapter 7 in regard to the physical built environment. This is not a limitation of the research or the Policies Index, but a limitation of the way in which CO_2 emissions are addressed in the policies.

This does not mean that at least some of the citizens and politicians of the 19 cities do not care about the issue of the location of emissions since this is being raised (C40 Cities Climate Leadership Group, 2022; Conci, 2021). The allocation of emissions is not simple. It is a very complex process that should include the origin of the product or service, extraction of raw materials, transportation either of the raw material or the product itself, recycling and disposal. This is simply the life cycle of everything that enters and is consumed in a city.

This raises issues about the high PIs in Table 8.18. High scores mean that the specific city's policies may be good pieces of legislation and may even suggest the city has good intentions to improve its environmental performance but this is not at all the same as doing it. The index itself, and the qualitative approach used in this research, also have limitations. For instance, the points given for the lobbyism criteria are highly subjective and based on legal lobbying. Aklin et al. (2014) have noted how corruption may affect the enforcement of environmental law through illegal lobbying. The PI is helpful in indicating the commitment of a city to the implementation of some sort of environmental impact control. However, it is unable to measure how efficient a city is in becoming sustainable. PI cannot indicate how sustainable a city is or will be, but only commitment to act towards some environmental impact mitigation. Some might consider it an assessment of how good a city is at "greenwashing".

Cities are actively pursuing a good environmental image, which is not the same as reducing environmental impact. Christos and Deffner (2005) identified the importance of city marketing to attract tourism, investment and business, all of which can be helped by having a green image. Furthermore, Wang (2019) has suggested that different stakeholder groups may interpret a city brand in different ways, and that tourists may think more of a green city brand than locals. Research commissioned by Nielsen Holdings, a global measurement and data analytics firm, using a sample of 30,000 people in all continents, showed that 81% of consumers considered eco-friendly practices as being very important for companies (Nielsen IQ, 2018), so it should be no surprise that cities seek a green brand through the policies they put into place. There is no way to be sure if a city's policies were originally developed to be a marketing tool, or if they were created to improve the environment. The fact is that regardless of their environmental efficiency, they can be used as marketing tools. Lucarelli (2018) has suggested that place branding is a specific form of urban policy.

The cities included in this research, and probably the majority of cities all around the world, face a similar scenario. They are keen to increase their reputation and attract more investment either via business or tourism. Consumers want to consume environmentally engaged outputs at any scale, from a small product to a city. Measuring the environmental impact and performance is very hard, particularly in large-scale areas. As a result, the use of environmental urban policies in any capacity is potentially a good initiative. Their effectiveness at achieving real strong sustainability is, however, at best challengeable. Lack of meaningful and trustworthy quantitative data and the difficulty of measuring real reductions in emissions lie at the core of the problem. The lack of standardised data makes environmental impact measurements and comparisons of city performances impossible. Perhaps the buzzwords are enough for the public and business perceptions, and perhaps such policies may lead to a meaningful improvement, but at this stage, it is impossible to confirm how effective the policies are in making the cities more sustainable. All the indications are that they are very far from enough and probably only scratch the surface of the problem. Part 2 of this book probes possible reasons for the current poor environmental performance of cities despite these supposed best intentions.

Note

1 Lea Hungrichhaußen and Fabiola Wimmer from Germany; Lea Medina and Hannah Bendull from the United States; Sai Lun Tin and Mingjie Qiu from China.

References

Aklin, M., Bayer, P., Harish, S. and Urpelainen, J. (2014). Who Blames Corruption for the Poor Enforcement of Environmental Laws? Survey Evidence from Brazil. *Environmental Economics and Policy Studies 16*(3), 241–262.

Bakshi, N. (2017). A Life Cycle Analysis of Living. PhD Thesis. Victoria University of Wellington.

Buenos Aires Ciudad (2022). Ecobici | Buenos Aires Ciudad. Retrieved from www.buenosaires.gob.ar/ecobici, 7 May 2022.

C40 Cities Climate Leadership Group (2020). Why Cities Need to Address the Construction Sector's Hidden Emissions. Retrieved from www.c40knowledgehub.org/s/article/Why-cities-need-to-address-the-construction-sectors-hidden-emissions?language=en_US, 9 May 2022.

Campbell, M. (2006). *Statistics at Square Two*. London: BMJ.

Chitturi, A. (2021). How Buenos Aires Is Transforming Its Mobility. Retrieved from www.hks.harvard.edu/climate/news-events/blog3, 29 December 2021.

Christos, L. and Deffner, A. (2005). City Marketing—A Significant Planning Tool for Urban Development in a Globalised Economy. 45th Congress of the European Regional Science Association: Land Use and Water Management in a Sustainable Network Society (pp. 1–21). Amsterdam: European Regional Science Association (ERSA).

City of Amsterdam. (2019). *New Amsterdam Climate*. Amsterdam: City of Amsterdam.

City of Amsterdam. (2021). Ambitions and Implementation Plan 2019. Retrieved from www.amsterdam.nl/en/policy/ambitions/healthy-sustainable/, 20 July 2021.

Conci, M. (2021). Let's Tackle the Carbon Emissions Hiding in Plain Sight: Neutralising Embodied Carbon in Buildings. Retrieved from www.climate-kic.org/opinion/how-can-we-leverage-space-innovation-to-take-climate-action-on-earth/, 9 May 2022.

Dalal-Clayton, B. and Bass, S. (2002). *Sustainable Development Strategies—A Resource Book*. London: Earthscan.

The Economist. (2021). *The Economist* Explains. Retrieved from www.economist.com/the-economist-explains/2013/09/02/is-the-economist-left-or-right-wing, 5 August 2021.

The Economist Intelligence Group. (2021). *Democracy Index 2020: In Sickness and in Health?* London: The Economist Intelligence Group.

European Commission. (2004). *National Sustainable Development Strategies in the European Union*. Brussels: EU. Retrieved from www.eunatur.de/attach/60/sustainable_development_strategies.pdf, 22 June 2021.

European Union. (2015). *Copenhagen—European Green Capital 2014*. Publications Office of the European Union.

Evans, B., Theobald, K., Joas, M. and Sundback, S. (2005). *Governing Sustainable Cities*. London: Earthscan.

Global Footprint Network. (2021a). Climate Change. Retrieved from www.footprintnetwork.org/our-work/climate-change/, 15 August 2021.

Global Footprint Network. (2021b). National Footprint and Biocapacity Accounts. 2021 Edition. San Francisco: Global Footprint Network.

Global Gridded Model of Carbon Footprints. (2021). Top 500 Cities Per Capita. Retrieved from http://citycarbonfootprints.info/, 26 July 2021.

Kern, K. and Gotlind, A. (2009). *Governing Climate Change in Cities: Modes of Urban Climate Governance in Multi-Level Systems*. London: Research Gate.

Krause, R., Feiock, R. and Hawkins, C. (2016). The Administrative Organization of Sustainability Within Local Government. *Journal of Public Administration Research and Theory 26*(1), 113–127.

Kurdziel, M-J., Nascimento, L. and Hagemann, M. (2020). Decreasing Costs of Renewables—Analysis of Energy Sector Planning and Climate Policy in Argentina. Retrieved from http://newclimate.org/publications/, 14 August 2021.

Lafferty, W., Knudsen, J. and Larsen, O. (2007). Pursuing Sustainable Development in Norway: The Challenge of Living Up to Brundtland at Home. *European Environment 17*(3), 177–188.

Lucarelli, A. (2018). Place Branding as Urban Policy: the (Im)Political Place Branding. *Cities 80*, 12–21.

Moran, D., Kanemoto, K., Jiborn, M., Wood, R., Többe, J. and Seto, K. (2021). Global Gridded Model of Carbon Footprints. Environmental Research Letters. Retrieved from http://citycarbonfootprints.info/, 26 July 2021.

Nascimento, L., Kurdziel, M-J., Fekete, H., Hagemann, M. and deVivero, G. (2020). Decreasing Costs of Renewables—Implications for Argentina's Climate Targets. Retrieved from http://newclimate.org/publications/, 14 August 2021.

New York City Council. (2021). Mayor's Office of Sustainability. Retrieved www1.nyc.gov/site/sustainability/index.page, 1 July 2021.

Nielsen IQ. (2018). Analysis. Retrieved from https://nielseniq.com/global/en/insights/analysis/2018/was-2018-the-year-of-the-influential-sustainable-consumer/, 13 December 2018.

Our World in Data. GDP World Bank. Retrieved from https://ourworldindata.org/grapher/gdp-per-capita-worldbank, 21 July 2021.

Pearce, F. (2021). Yale Environment 360. Retrieved from https://e360.yale.edu/features/paris-conundrum-how-to-know-how-much-carbon-is-being-emitted, 12 August 2021.

Sühlsen, K. and Hisschemöller, M. (2014). Lobbying the "Energiewende". Assessing the Effectiveness of Strategies to Promote the Renewable Energy Business in Germany. *Energy Policy 69*, 316–325.

UEA. (2021). Dubai Clean Energy Strategy. Retrieved from https://u.ae/en/about-the-uae/strategies-initiatives-and-awards/local-governments-strategies-and-plans/dubai-clean-energy-strategy, 3 July 2021.

United Nations (2021a). United Nations Department of Economic and Social Affairs. Retrieved from www.un.org/en/development/desa/population/publications/urbanization/urban-rural.asp, 1 June 2021.

United Nations (2021b). United Nations Development Programme. Retrieved from http://hdr.undp.org/en/content/human-development-index-hdi, 20 July 2021.

Volkery, A., Jacob, K., Bregha, F., Pinter, L. and Swanson, D. (2004). Coordination, Challenges and Innovations in National Sustainable Development Strategies. Berlin Conference on the Human Dimension of Global Environmental Change "Greening of Policies: Inter-linkages and Policy Integration". Berlin: IISD.

Wang, H.-J. (2019). Green City Branding: Perceptions of Multiple Stakeholders. *Journal of Product and Brand Management 28*(3), 376–390.

PART 2

Influences on living in cities

9

ENVIRONMENTAL DEATH BY DEMOCRACY

9.1 Introduction

Although we all think we know what it means, the definition of democracy is elusive and has no consensus; it has a variety of potential meanings and is complex to define (Dalton et al., 2007). However, nearly 80 years ago Schumpeter (1943) provided a broad definition that may help to cluster countries according to their political regime. He described democracy as a regime holding contested elections—contestation and participation. An in-depth explanation of democracy, or the concept of democracy, or of a democratic state is not the aim of this chapter, but some relevant aspects and common characteristics of democracies in different countries will be discussed.

Nevertheless, the title of this chapter and the subtitle of this book may lead to some misunderstandings. This book does not aim to assess the success of democratic governance in different countries or what those countries have achieved in terms of development. The aim here is to understand how and why democracies are apparently failing in addressing environmental impact. As a result, in this chapter, there is no in-depth comparison between different types of democracies and how they handle environmental problems, nor talk about left- or right-wing politics, liberal or welfare states and their achievements, or not, in fighting climate change. We will present an overview of how what are perceived as democracies are handling the current challenging environmental situation. Common aspects of democracies have been identified. The potential failure to reduce environmental impact in democracies does not necessarily mean that an autocratic system or a literal dictatorship is better in addressing environmental matters. It may simply mean that democracies, as good as they may be in other fields, are unequipped to handle our current environmental situation. It is easy to see that democracies are responsible for more significant environmental impact than some non-democratic countries,

DOI: 10.4324/9781003045847-11

although assuming this is simply the outcome of increased development is a superficial assessment of the current world environmental situation. One of the main points of this chapter is to understand why, as democracy depends on votes, voters are often not motivated to vote in those countries that have a clear environmental agenda, or why voters do not demand more representatives with strong environmental commitments.

9.2 Why are democracies not working to stop our environmental catastrophe?

Although there is no consensus on the proper definition of democracy, there are some similarities in countries perceived as democratic. As there are many types of democracies (Schimitter and Karl, 1991), the perception of what a democracy is will differ among individuals. Dahl (1971) described some relevant aspects necessary for a democracy. These included freedom to form and join organisations, freedom to vote, access to alternative sources of information and free and fair elections. According to Dahl (1971:3), when citizens can join equally in free and fair elections, and if elections guide the government's actions, then the standards of democracy are met. Schimitter and Karl (1991) suggested that modern political democracy is a system of governance in which rulers are held accountable for their errors in the public sphere by society, acting through the competition and cooperation of their elected representatives. However, de Mesquita and Smith (2011) have identified that no two democracies or dictatorships are alike. They present a clear way to explain one of the significant differences between democracies and dictatorships—while democracies rely on a large number of people to make some decisions, dictatorships rely on a much smaller group of people.

As hard as defining democracy may be, classifying democracies is equally challenging. Boese (2019) compared strengths and weaknesses across the three most prominent democracy datasets: Polity IV, Freedom House, and Varieties of Democracy. Her research suggested that Varieties of Democracy (V-Dem) outperforms the other two. The criteria are different in the three datasets and the results are used for different purposes. The Freedom House Institute was designed to measure civil liberties and political rights. Polity was conceived for analysing the durability of institutional frameworks, while the Varieties of Democracies is situated somewhere in between (Boese, 2019). Similar research was carried out by Elff and Ziaja (2018), with similar results. Despite Boese's efforts to compare the datasets and the resulting indices, and determine the most effective one, and Elff and Ziaja's support for V-Dem, our research has used the Democracy Index from the Economist Intelligence Unit, which was not included in Boese's research. In Hungrichhaußen's (2021) research into energy and transportation policies, which was the base for Chapter 8 on urban policies, the index used was from The Economist Intelligence Unit (EIU). As a result, to maintain consistency, the same index has been used in this chapter.

Because the findings of Boese's and Elff and Ziaja's research point to the usefulness of a different index, a correlation between their index (V-Dem) and the one selected here was made. The Democracy Index from EIU evaluates the electoral process and pluralism, the functioning of government, political participation, political culture and civil liberties, while V-Dem has data grouped into components of democracies: liberal, egalitarian, participatory and deliberative (V-Dem, 2022). As this chapter does not aim to compare democracies or measure the effectiveness of indices, the strong correlation between V-Dem and the Democracy Index (DI) from EIU ($r > 0.9$) was considered enough to support the indicator chosen for this chapter and keep the research consistent.

The EIU divides countries into four categories according to the overall score—full democracies, flawed democracies, hybrid regimes and authoritarian regimes. Scores are given from 0 to 10, with scores between 10 and 8 classified as full democracies, between 6 and 7.99 flawed democracies, between 5.99 and 3 hybrid regimes, and below 2.99 authoritarian.

9.3 Assessing the relationship between democracies and the environment

The comparison between countries, their governmental system and their environmental performance in this chapter was primarily based on the relationship between the EIU DI and the Ecological Footprint (EF) in global hectares (gha) from Global Footprint Network (GFN). However, income per capita and the United Nations Human Development Index (HDI) were also considered in an attempt to find other relationships between systems of government and environmental impact.

The countries were selected based on the availability of data from GFN for calculation of EF in gha. The correlation between the DI and environmental impact was calculated including all countries for which data was available from both the EIU and GFN.[1] Using the categories in the DI, the total population in each category along with the average EF, HDI and income per capita were calculated (Table 9.1). Population, EF, HDI and income average are from the Ecological Footprint dataset from The Global Footprint Network (2021).

There are interesting aspects to be seen in Table 9.1. While the correlation is not particularly strong, there is a suggestion that a high democracy index leads to a high environmental impact. Arguably, it is the result of high development as can be seen by the high income and HDI for the Full Democracy segment. This raises the issue

TABLE 9.1 Comparison of government types with EF, HDI and income per capita

Government system	*Total population in each system*	*EF average (gha)*	*HDI average*	*Income per capita average ($US)*
Full democracy	456,044,000	5.3	0.921	46,713.20
Flawed democracy	2,982,866,000	2.7	0.729	11,563.87
Hybrid regime	1,243,680,000	1.5	0.613	3,429.88
Authoritarian	2,802,260,000	3.0	0.688	5,844.73

that neither DI nor HDI include assessment of environmental impact. The other democracy index, V-dem, also excludes environmental impact in their assessment. Therefore, as environmental impact is not included as a criterion, it is fair to assume that a high HDI may be achieved even if the country is highly unsustainable. This unsustainability may include cheap and polluting energy sources (often from outside the country's borders), deforestation and exploitation of natural resources, and trade with environmentally deregulated countries.

Table 9.1 suggests some complex figures and less obvious facts. These help to clarify why there is no strong correlation between DI for all the countries and their environmental impact. As can be seen, authoritarian countries have higher HDI and environmental impact than hybrid regimes. Of the 33 countries used to calculate the average values for hybrid regimes, 18 are in Africa, considered to be "the poorest continent on Earth" (SOS Children's Villages, n.d.). This partially explains the lower HDI and income per capita in this DI group. Other reasons for possible distortion include the fact that 13 of 15 countries in the Organization of the Petroleum Exporting Countries (OPEC) are categorised as authoritarian. Some of these countries have higher HDIs and higher EFs than some countries in the higher tiers of the DI.

An understanding of the DI and an overview of how modern democracies often operate is essential for assessing their capacity to handle the environmental challenges ahead. Samet and Schmeidler (2003) identified the apparent dominance of liberal democracies in the world. More recently, Bell and Building (2014) suggested that democracy occurs because it produces a reasonably comfortable life. The population figures in Table 9.1 show that democracies, both full and flawed, are in the minority. It is important to note that democracy is often linked with the concepts of capitalism and liberalism (see Chapter 10). Early on, Schumpeter (1943) identified the link between democracy and capitalism and suggested that modern democracy arose along with capitalism. In a more recent association, Russell et al. (2007) have identified the connection between liberalism and democracy. Phelan and Dawes (2018) defined liberalism as a political philosophy that privileges individual liberty, property rights and market freedom over mercantilist trade restrictions. Although the philosophical discussion around liberalism is not the central part of this chapter, this short definition is enough to suggest the modus operandi of democracies and the importance to many of them of the free-market approach. An in-depth exploration of the definitions of neoliberalism, which has been defined as a revival of liberalism (Thorsen and Lie, 2007), is not necessary for the full understanding of the main aspect of this investigation. Chomsky (1999) suggested that neoliberalism was the spread of global consumerism and the demolition of the proactive welfare state, and that finally it would aim to scrap environmental regulations. To understand the concepts of liberalism and neoliberalism clearly, it is necessary to identify the relationship between them, capitalism and democracy (Mudge, 2008). Liberalism comes back to John Locke's (1632–1704) approach to private property (Bell and Building, 2014; Samet and Schmeidler, 2003; Stephens, 2001). Locke's argument includes the concept that unused land was essentially waste (Harding, 2019), preparing the conceptual justification for private property and exploitation of natural resources. The concept of the free market, the rights of human beings over nature and private property are aligned with the historical appearance of the idea of free-market capitalism and Western-style representative democracy. Neoliberalism is more recent and goes even further in the relationship between economy and politics. Neoliberalism is a type of liberalism, capitalism and democracy that has been globalised since the 1970s, and potentialised by Ronald Reagan and Margaret Thatcher (Thorsen and Lie, 2007). The importance of the free market for the current concept of democracy is a relevant part of this discussion. The free market itself is not the problem, but capitalism's structure and its permanent need to grow is a problem for the environment (see Chapter 10). It seems that the relationship between democracy and liberalism serves as a mechanism for maintaining economic growth. This idea is not unique or even new. Canclini (1995) suggested that the exercise of citizenship is directly associated with the capacity to appropriate assets; going beyond the right to vote or to form a union. It is surprising that this relationship is not more often commented on as, for example, in the Human Development Index (HDI) from the United Nations (2022a). As stated earlier, this measures life expectancy, years of schooling and a decent standard of living, measured by gross national income (GNI) per capita. In democracies, citizens exercise their right to vote and their right to consume. An election candidate who offers voters less consumption, in order to deal with environmental issues, is unlikely to win against a candidate who offers more.

8.4 The power of the majority, the power of voting

Schmitter and Karl (1991) indicated that democracies identify themselves with majority rule, meaning that any government makes decisions based on the combined votes of more than half the voters. This majority may emerge within an electorate, a parliament or a city council. Democracy relies on voters voicing, through voting, their preferences in regard to policies. Citizens have the right to vote (not everyone is always allowed to vote, and in some cases, it is not a right, it is an obligation). The representative, a person, a party or a coalition, who gains more votes and so represents the majority, is entitled to represent the entire electorate. There are two major issues with this process as it relates to the environment (and perhaps also with many other issues). The first problem lies in the representation process itself. This is based on convincing the majority of the voters that a certain future representative will implement a group of policies. During this process of convincing voters of intent, it is not feasible to cover all possible policy aspects. There may be documents covering issues in greater depth than in the electoral campaign, but it is impossible to cover all the issues in the parties' platforms, particularly because new unpredictable issues may occur during a representative's term. This means that voters have to trust their representatives to decide on their behalf even on issues that could not be predicted, or on items of which they have no knowledge. For environmental problems, such as climate change, a meaningful debate and decisions about these items would require scientific knowledge. At the same time, it is not expected that a member of parliament would be an expert on climate change. The voter has relied on a representative, who possibly will rely on advice from an expert and take a decision that may, or may not, be the same as the decision of those that voted for them. Although there are ways to verify what the electorate thinks about a specific issue, such as focus groups or surveys, it is not as representative as the electoral process itself as it does not cover the same share of the population. Referenda could be a way to legitimise all decisions, but it would be a more expensive and slower process. Although elections are the forums where politicians are judged, there is little or no way of making politicians accountable, as long as the decisions made were legal.

This raises another important aspect related to the legality of decisions taken by politicians. In a presidential democracy this would be the president, or a prime minister in a parliamentary democracy. Either person might make terrible decisions related to the environment, but the power invested in them makes a bad decision not only executable but totally legal since they have been democratically elected. There are several ways to stop a bad law, but it is possible this may take a lot of time, and perhaps the issue will be irrecoverable. At the end of the elected term, and if the electorate is well informed, the representative may be replaced through the vote, although this does not mean that the environmental impact will be averted. Although hypothetical, this process is both legal and plausible. However, environmental issues do not respect election cycles. A bad policy may be irreversible and the next election may be too late to change some bad decisions. There are mechanisms for changing governments between elections, but they are rare, and rarer when related to environmental issues; we could find none at the national level. This makes changes in democratic systems by definition slow, particularly if considered against the emergency of the current environmental situation.

The democratic system is based on the majority of people electing a representative who will govern on their behalf, making decisions on things for which they do not have expertise. To do this they need to rely on someone who needs to be unbiased and capable of supplying complex information on which non-experts can make decisions. The majority representative needs to be able to identify unbiased experts and persuade them to provide advice for their decisions. Vestheim (2012) has suggested that all the players involved in this process do not act from a neutral point of view. This is probably a simplification of the entire process, and there are likely to be many processes for obtaining the best possible advice in fields where the representative is not an expert. If something goes wrong in this scientific information chain, it is likely the only way to fix the problem will happen after a new election, potentially in some years. In addition, it is quite possible that a government, faced by economic and other pressures, will not act to prioritise the environment, which is a long-term issue, rather, they will favour short-term decisions in the hope of gaining re-election.

The nature of this process of gathering and organising information for government decisions is complex and opens the possibility for external players to join the decision-making process. Lobbying is one of the ways that external parties can influence politicians during the periods between elections (Bernhagen, 2013). People can make their voices heard between elections through public audiences, protests, strikes and more recently through social media (see Chapter 12). Companies can buy advertising space in traditional media, social media or they can lobby politicians. All people's representatives are subject to external influences. The level of influence may be legal (advertising and lobbying) or illegal (bribes). The complexity lies in the influence of lobbying, which may be legal but not entirely ethical. The entire character of the lobbying process has been regarded as a potentially corrupting influence on politics (Blach-Ørsten et al., 2017; Moloney and McGrath, 2019).

The act of influencing decisions with a negative impact on the environment, which are against the collective interest, does not happen exclusively inside a governing body. In 2010, the U.S. Supreme Court held, in the case called Citizens United against the Federal Election Commission, that free speech in the First Amendment of the American Constitution forbids the government from limiting independent expenditures for political campaigns by companies, unions and non-governmental organisations (Gesami, 2021; Levitt, 2011; Teachout, 2010). This means that in the U.S. companies are legally entitled to buy as much advertising as they can afford to convince the electorate that, for example, climate change is not a problem. In this case, if the advertising works, it is likely that voters will swing towards those that do not defend environmental causes. It appears free speech in the United States is free only for those who can pay for it.

The entire representative system on which democracies are based seems very fragile when it needs to act rapidly to reduce the effects of climate change and environmental impact. This is not surprising. Changes require a majority, and establishing the majority either during an election or inside the governing body may take too long or may never be achieved. In a political process to achieve a majority, different players will act, including individuals and companies. While individuals may have a vested interest in reducing environmental impact, companies do not necessarily have this. People normally manifest interest in things that affect themselves and their community (Sheppard et al., 2011), while companies, particularly large ones, have a diversified ownership as their stocks are dealt with everywhere in the world, so they may lack any sense of community, and their focus is only on the money they can make. If the interests of companies and people are different, companies are likely to oppose and slow down any process of change. Companies can, for example, fight to postpone a certain law or regulation. However, both people and companies often share interests, such as economic growth. This common interest is particularly clear given only nine countries in the world have not had a real increase in their gross domestic product (GDP) since 1950, and on average, the world is 4.4 times richer (Our World in Data, 2022). Perhaps, for this reason, the desire for economic growth unites the fragmented social spectrum, often composing the majority and fuelling a strong political-economic structure (Akbulut, 2019). Economic growth, measured through GDP, has been widely misused internationally for over a half-century to measure well-being and prosperity (Costanza et al., 2018). Thus, it is understandable that economic growth is an essential part of people's political agenda, meaning more consumption, which is good for companies. As a result, it is easy to see how companies and people unite on this political topic. The environmental performance established by the Paris Agreement may provide some indication of how countries are pursuing their environmental goals, particularly the democratic ones, but as economic growth is a uniting factor, it is conceivable that the majority of governments are formed around topics related to economic growth rather than environmental goals.

Besides being potentially slow, the process to gain a majority so as to achieve changes in a democratic system does not necessarily mean that the majority will make the right decision. The achievement of a majority does not define what is right or what is the truth, it is an outcome of competition, especially when the matter is complex, and the decision-makers are not experts in that matter. This again brings up again the role of lobbies. Lobbying does not work in the same way everywhere. In some places, there is specific legislation regulating political lobbying. For example, lobbying in Australia has a code of conduct and a public register of lobbyists (ANZSOG, 2021). In Denmark, 11% of the former members of parliament have turned to professions linked with lobbying (Blach-Ørsten et al., 2017). In a speech in 2003 in New Zealand, Trevor Mallard (currently Speaker of the New Zealand Parliament) said that lobbying was a legitimate activity in a democracy (Labour Party, 2022; New Zealand Government, 2022). Currently, a bill for lobbying transparency is being discussed in the New Zealand Parliament (New Zealand Parliament, 2022). Countries like Brazil and Argentina do not have proper legislation to define the limits for lobbying (Verly and Ruiz, 2022; Wilson Center, 2022). Even in a democratic decision-making process where there are no lobbies, a more charismatic orator may be able to defend an environmentally unfair proposition and attract votes. Subedi and Scott (2021) note that rather than charisma being a personal inherent quality, it is socially and politically developed and is a kind of political performance. This suggests that it is possible that in defending commercial interests, companies can develop a complete strategy that spans from advertisements, the use of "junk science" and parliamentary lobbying as far as promoting a charismatic politician to convince the electorate.

The full democratic structure, based on convincing people to support a certain cause or person, seems to be contributing to the current inaction towards the environment and climate change. Shwed and Bearman (2010), when trying to explain what scientific consensus means, have put forward some intriguing questions regarding how societies' perceptions change following scientific discovery or consensus. For instance, when and why has society generally accepted that smoking causes cancer? They argue that it is based on the achievement of certain levels of scientific consensus. However, while smoking is decreasing worldwide (Our World in Data, 2019c), the environmental impact of greenhouse emissions is increasing

(Friedlingstein et al., 2020) despite cancer from smoking and anthropogenic climate change having a similar level of scientific consensus (Shwed and Bearman, 2010). There are several factors that may explain the difference between smoking and climate change. For example, widespread research related to smoking problems goes back much further to the 1950s, against the early 1980s for anthropogenic climate change (Shwed and Bearman, 2010). Even extremely powerful and transnational lobbying has not been able to keep smoking up and it is falling worldwide (Amul et al., 2021; Our World in Data, 2019c; Watts et al., 2020). A major difference between the smoking and climate change lobbies is clear. While smoking is a very large industry, it is still a single industry sector, the tobacco industry. Therefore, it is a single industry fighting against the scientific consensus. In contrast, a large-scale scientifically based climate change policy may directly affect consumption, which means challenging not just a single industry but the very core of how modern society works and defines itself (Campbell, 2018). Moreover, it will require people to relinquish their familiar, comfortable modes of living (Jasanoff, 2010) and will affect not only one industrial sector but potentially all the consumption system, meaning the whole industrial sector, as well as individuals' ways of life and sense of personal value. The development of policies or laws that ultimately reduce environmental impact, in theory, targets consumption by either reducing it or making prices higher. An increase in prices could be through the application of environmental taxes or making production more complex and therefore more expensive. If the principle of reducing environmental impact means reducing consumption, which affects corporations' interests, they will want to convince people that environmental impact is not a real issue or that certain products are environmentally friendly. If people perceive a product as eco-friendly, when in fact it is harmful, or think that climate change is a problem made up by scientists, it is possible that these people will not have the environment as the top priority when they choose their representatives. In fact, voters do not base their choice on scientific facts, they base it on their perception of the facts (Pearce et al., 2017). If the voter does not even perceive the problem, their representative may well not act to promote pro-environmental policies.

Political inaction is not only the outcome of lobbying or the efforts of industries. Inaction towards environmental impact and climate change has been studied using the psychological construal level theory (CLT), which assesses how individuals engage with events. CLT, in a very simple way, suggests that closer events are perceived in a more concrete way while distant events are considered more abstractly (Maglio et al., 2013). The event distance is studied in four different dimensions—social, hypothetical, spatial and temporal (Trope and Liberman, 2010). Social refers to who is affected while hypothetical is related to the likelihood of a certain event happening. Spatial deals with the location and temporal when it will happen (Spence et al., 2012). Environmental impact is largely perceived as a distant and abstract problem (Chu and Yang, 2020). Jones et al. (2017) suggested that people perceive environmental impact as something distant, and as the perceived distance increases, the event (in this case environmental impact or climate change) becomes more abstract and decontextualised. Through applying CLT to environmental impact, surveys indicate that it is part of human nature to perceive environmental issues as something more abstract, decontextualised and potentially less urgent than many other aspects of daily life. For example, 57% of Americans believe that global warming will not pose a serious threat to their way of life in their lifetime (Gallup, 2021). This figure suggests that all four of the distance dimensions from CLT may be seen among environmentally sceptical American citizens. It is fair to assume that some respondents may think it will happen to someone somewhere else (socially and spatially), it is unlikely to happen to me (hypothetically), or it will happen after my lifetime (temporally). It is also fair to believe that similar thoughts might pass through human minds anywhere in the world. A person's own experience of the nature of environmental impact does not automatically trigger the urgency to act against environmental degradation. The temporal dimension of the CLT plays another relevant role in the democratic arena, this being the difficulty that democratic systems have to provide a significant representation of future generations. The Intergovernmental Panel on Climate Change (IPCC) have suggested that current emissions would lead to median global warming of 3.2°C (2.2–3.5°C) by 2100 (United Nations, 2022b). Mitigating this will require actions that go beyond not only the next election but the lifetimes of politicians, it means legislating for people that have not yet been born. Thus, it is hard to create a majority around topics that are not perceived as either urgent or personally threatening.

Hence in democracies addressing environmental issues, environmental impact may not be perceived as a concrete problem, politicians are often not educated in science, political decisions may be legal but not environmentally focussed, and industry lobbying may aim for increasing consumption along with people's desire to achieve or keep a certain social level (Jasanoff, 2010); all of these issues make it hard for democratic systems to make meaningful changes. People relate consumption to higher social levels (Iyer and Muncy, 2016), therefore giving up consumption may be perceived as social status regression. This is validated by the philosophical system justification theory (SJT) (Feygina et al., 2009), which suggests that societies need to maintain a sense of certainty and stability. As a result, criticism of the lack of the capacity of democratic systems to address environmental problems goes against the natural motivation to perceive the system as legitimate and able to handle complex and vast changes. In theory this makes democratic systems unprepared to make radical changes to address issues that challenge many of the democratic principles. Essentially, SJT suggests that it is normal for people and a society to try to maintain and protect the status quo.

Majority rule, along with equality of citizens and widely inclusive citizenship, is a key factor in democracies (Galston, 2018). According to the democratic principle, hypothetically the majority in a country could decide to drop every pro-environment law. Although it would be democratically acceptable, it would not be acceptable under environmental criteria and would not be a decision based on scientific principles. The point is that the majority decision, one of the main pillars of democracy, is not beholden to any scientific fact. When former US President, Donald Trump, who was democratically elected in the United States, withdrew from the Paris Agreement on climate change in 2020 (BBC News, 2020) the fact that he had been elected to do this did not make the act more legitimate from an environmental point of view.

The IPCC (2022) states, with high confidence in their data assessment, that a global temperature rise of more than 1.5°C by 2100 would generate a significant loss of ecosystems. Even limiting the warming to 1.5°C means that sea-level rise will continue beyond 2100. In order to keep warming below 2°C, and aiming to keep global warming below 1.5°C, immediate actions are required (IPCC, 2022). Countries that signed up to the Paris Agreement in 2015 agreed to cut greenhouse gas emissions with a view to "holding the increase in the global average temperature to well below 2°C above pre-industrial levels and pursuing efforts to limit the temperature increase to 1.5°C above pre-industrial levels" (IPCC, 2017). Considering that countries joined the Paris Agreement voluntarily, their environmental performance in reducing their emissions could be a good indicator of their commitment to improving their environmental impact. An independent scientific analysis produced by two research organisations (Climate Analytics and NewClimate Institute) tracks climate actions to achieve the Paris Agreement goals in the Climate Action Tracker (CAT). This covers 39 countries and the European Union or 85% of the total global CO_2 emissions (Climate Action Tracker, 2022). The CAT has divided countries into different groups according to their performance against their own goals, assessing mitigation targets, policies and action. The countries' efforts are classified as Critically Insufficient (CI), Highly Insufficient (HI), Insufficient (I), Almost Sufficient (AS) and Compatible (C). Table 9.2 shows the performance of all countries included in the CAT assessment but separated according to their political regimes based on the DI (The Economist Intelligence Group, 2021), together with the national EFs (Global Footprint Network, 2021) and average EF for the group.

The first thing to note is, that regardless of the political regime, no country is reaching what is considered compatible with the goals agreed upon during the Paris Agreement. However, looking at the two categories of democracies, full and flawed, only two countries (the UK and Costa Rica) are classified as AS (Almost Sufficient). All others, including all countries in the European Union, are classified as I (Insufficient) or HI (Highly Insufficient). Of course, each country has their specific reason for not producing results compatible with the Paris Agreement, but the fact remains that there is no democracy in the world committed to the points agreed upon during the Paris Agreement Summit. Table 9.2 suggests that even highly democratic countries are acting slowly or not acting at all to minimise their environmental impact. The last column shows the Ecological Footprint of each nation from the Global Footprint Network (2021). Considering a sustainable nation would have an EF under 1.59 gha/person, only India, Nepal, Gambia, Kenya and Ethiopia would be considered sustainable under the GFN assessment, and of these

TABLE 9.2 Climate mitigation action, DI and EF

Classification according Democracy Index from The Economist Group 2021	*Countries*	*Status of each country according to CAT assessment*	*Ecological Footprint per-capita annum (gha)*	*Ecological Footprint average (gha)*
Full democracy	South Korea	HI	6.17	5.87
	Canada	HI	8.07	
	*Australia**	HI	7.27	
	*New Zealand**	HI	4.32	
	Austria	I	6.02	
	Japan	I	4.65	
	*Germany**	I	4.70	
	Luxembourg	I	12.79	
	*Netherlands**	I	5.02	
	Switzerland	I	4.47	
	Ireland	I	5.01	
	*Denmark**	I	6.92	
	Sweden	I	6.06	
	Finland	I	5.83	
	Norway	I	5.77	
	Costa Rica	AS	2.56	
	United Kingdom	AS	4.20	
Flawed democracies	Thailand	CI	2.57	4.25
	*Singapore**	CI	5.86	
	Colombia	HI	1.91	
	*Indonesia**	HI	1.66	
	*Argentina**	HI	3.30	
	*Brazil**	HI	2.80	
	*India**	HI	1.19	
	Peru	I	2.14	
	Romania	I	3.40	
	Croatia	I	3.71	
	Hungary	I	3.67	
	Bulgaria	I	3.63	
	Poland	I	4.71	
	Slovakia	I	4.41	
	South Africa	I	3.15	
	Lithuania	I	5.86	
	Latvia	I	6.13	
	Cyprus	I	3.87	
	Belgium	I	6.60	
	Slovenia	I	4.90	
	Greece	I	4.12	
	Malta	I	5.68	
	Italy	I	4.41	
	Czech Republic	I	5.47	
	Portugal	I	4.40	
	Estonia	I	7.16	

TABLE 9.2 Cont.

Classification according Democracy Index from The Economist Group 2021	*Countries*	*Status of each country according to CAT assessment*	*Ecological Footprint per-capita annum (gha)*	*Ecological Footprint average (gha)*
	United States of America★	I	8.03	
	Chile	I	4.28	
	Spain	I	4.02	
	France	I	4.60	
Hybrid systems	Mexico	HI	2.62	1.82
	Turkey	CI	3.50	
	Gambia	AS	0.93	
	Nepal	AS	1.09	
	Kenya★	AS	0.99	
Authoritarian regime	Saudi Arabia	HI	5.76	4.23
	China★	HI	3.71	
	The United Arab Emirates★	HI	8.94	
	Egypt★	HI	1.78	
	Kazakhstan	HI	5.98	
	Iran	CI	3.22	
	Viet Nam	CI	2.24	
	Russian Federation★	CI	5.47	
	Ethiopia	AS	1.00	

★ Countries in which the 19 cities are situated.

only India is HI, the other four being AS (Almost Sufficient). It seems that if you are not a highly developed nation, it is easier to meet the Paris Agreement target, which somewhat puts the developed nations with all their resources to shame. The Ecological Footprints in Table 9.2 also indicate that besides not doing enough to achieve the Paris Agreement target, democratic nations are using more resources and causing more environmental impact than non-democratic nations. A link can be seen between HDI (Table 9.1) and democratic systems, and the same connection can be seen between their environmental impact and lack of demonstrated intention of sorting this out. However, in 2019 the two full democracies scoring AS (UK and Costa Rica) both fall into the very high HDI category (above 0.8), Nepal and Kenya fall into the medium HDI category (both 0.6) and Gambia and Ethiopia into the low HDI group (both 0.49) (numbers rounded) (UNDP, n.d.). Costa Rica with a low EF of 2.56 gha, a very high HDI and a score of AS for meeting the Paris Accord looks like a country to study further, but all may not be as good as it seems. The change of elected president in 2022 has been partly put down to public reaction to the high cost of living and issues with crime. "Costa Rica has enjoyed relative democratic stability compared with other countries in the region, but the public has grown frustrated with public corruption scandals and high unemployment" (Cordoba, 2022).

8.5 Will democracies' pursuit of progress lead to environmental sustainability?

Kasman and Duman (2015) have argued that both existing countries from the European Union and those that are candidates to join the European Union may present an inverted U-shaped curve in a graph comparing development and environmental impact. This is called the environmental Kuznets curve hypothesis (EKC). The EKC suggests that after reaching a certain development level (tipping point), countries start reducing their environmental impact. This environmental assessment is a derivative of the original study by Simon Kuznets who hypothesised that income inequality rises, reaches a tipping point, and then reduces as income per capita increases (Kuznets, 1955). Similarly, Chen et al. (2019) suggest that China may reach an inverted U shape if a strong policy of renewable energy is applied. Choi et al. (2010) compared China, South Korea and Japan and found that Japan, the most developed nation in their sample, had reached the peak and had an inverted U-shaped EKC. The EKC would provide a perfect justification for pursuing prosperity through economic growth, as if the hypothesis were correct the countries that have not yet achieved a certain development level (the tipping point) could keep their goals of increasing economic growth. On this assumption, the democratic machinery with lobbies and advertising and the majority ruling would be a good response to the environmental problem in countries that have not yet achieved their development goals. This assumes that consumerism will bring environmental development after reaching a certain level of consumption. There are, however, some interesting points related to the EKC. First, the original economic hypothesis from Simon Kuznets was subjected to a large study, which pointed out the weakness of the theory (Piketty, 2017). Piketty argued that the original Kuznets theory was based on a very small and special time frame, in which the world economy was being rearranged post World War II. The idea that developed nations are able to reduce their environmental impact and emissions without meaningfully reducing consumption seems at the best fragile. While developed nations have reduced their emissions, nations like China and more recently India have dramatically increased theirs (Our World in Data, 2019a). At first sight, this seems to support the EKC theory, but the data comparing the participation of industry's share in the economies of countries shows that industries have moved their production and emissions from democratic and developed nations to China and more recently India (Our World in Data, 2015), while the consumption of the products of those industries is still heavily concentrated in democratic and developed nations (Our World in Data, 2019b). Extensive academic literature points out the limitations of the EKC (Cole et al.,

2001; Dasgupta et al., 2002; Dinda, 2004; Karsch, 2019; Nahman and Antrobus, 2005; Stern, 2004). These studies indicate that the EKC cannot be verified and the benefits are normally limited to local pollutants rather than global-scale ones such as carbon dioxide. The importance of the EKC lies in people's perception. The EKC legitimates the democratic effort to achieve prosperity, because the theory says this will lead to better environmental performance.

8.6 Discussion and conclusion

The democratic process requires people to believe and like someone and their ideas, casting their votes for these ideas as represented by parties or candidates. However, environmental laws may not be popular at the hustings. They may limit consumption, increase prices, ban products or simply change people's lifestyle. This raises questions regarding how the citizens of democracies would react if, following an election, unpopular laws were imposed against the wishes of the majority of people. What would happen if a government decided to limit people's capacity to move around in their own cars (1970s' petrol rationing), limit international air travel (2019–2022 COVID border lockdowns), or even limit meat eating (World War II food rationing)? These are all things that governments have done because they have had to be done in the interest of national survival. However, before an election, would a politician put forward any of these propositions, undeniably good for the environment, but likely to be unpopular and possibly not beneficial to the economy? Ultimately, the majority legitimates the unsustainable actions of democratic governments, as they decide who is to govern. This brings up another issue regarding the majority representation democratic system; different species, ecosystems and future generations are not fully represented. Although some may argue that different species and ecosystems do not have the same rights as us, the same cannot be said of future generations. An individual born in 2022 could well be alive in 2100 and will live to face all the consequences of our emissions and our omissions, which means that generation 2100 is already alive but helpless.

Democracies rely on the balance and independence of power between different government branches, in general, divided into legislative (Parliament), executive (Cabinet and Ministers outside Cabinet, plus government departments) and judiciary. The principle of checks and balances states that separate branches are empowered to prevent actions by other branches and are induced to share power (Britannica, 2019). Although there are relevant differences between presidential and parliamentary systems, they are essentially the same; the independence between the branches aims to allow them to check each other and prevent any of them from becoming excessively powerful (Cornell Law School, 2022). Governments will create their own mechanisms, normally through laws or regulatory agencies, to handle environmental issues, but even if the balances and checks of the democratic system work for a wide range of topics, environmental impact reduction will always rely on the quality of legislation, which is susceptible to all the democratic problems discussed above.

This chapter does not strive to find a solution to the inability of democracies to handle the environment but rather to point out the issues involved. Perhaps the representation of environmental interest, which includes future generations, different species and ecosystems, should be subjugated to a special branch of government, equally independent and based on science so as to able to assess environmental viability. However, if this branch were susceptible to the usual system of checks and balances, its findings and policies could be over-ridden by other branches of government for political, economic or social reasons. The checks and balances inevitably make democracies more complex and often inefficient in addressing urgent matters, even assuming that voters will support environmental policies. Discussions about degrowth may provide a different possibility with a new pathway for a more sustainable future, but it will shake the principles of democratic systems. In fact, Pichler et al. (2018) have pointed out the capitalist and fossil-fuel-based foundations of the democratic states and the relationship between these states and the environmental crisis. The social transformation required in order to ensure a less disastrous environmental future is enormous (Brand, 2016; Görg et al., 2017; O'Brien, 2011; Pichler et al., 2017). Currently, neither democracies nor dictatorships have chosen to address the issues that our society is facing in terms of environmental degradation and climate change as described and agreed on in the Paris Agreement (Climate Action Tracker, 2022). A positive note came from the pandemic. Several nations were capable of quickly adapting and imposing lockdowns and as a result reducing consumption (see Chapter 14). The fact remains that if we perceive danger, we are able to act fast. The trouble is perhaps that in a world run by money we are not allowed to perceive the danger.

Note

1 *Full democracies*: Costa Rica, Mauritius, United Kingdom, Japan, Republic of Korea, Germany, Luxembourg, Canada, Netherlands, Australia, Switzerland, Ireland, Denmark, Sweden, Finland, New Zealand and Norway; *Flawed democracies*: Montenegro, Republic of North Macedonia, Thailand, Peru, Republic of Moldova, Papua New Guinea, Albania, Sri Lanka, Singapore, Guyana, Lesotho, Serbia, Mongolia, Romania, Republic Dominican, Colombia, Croatia, Ghana, Hungary, Namibia, Philippines, Bulgaria, Indonesia, Poland, Argentina, Suriname, Panama, Brazil, India, Slovakia, South Africa, Timor-Leste, Jamaica, Trinidad and Tobago, Lithuania, Malaysia, Latvia, Cyprus, Belgium, Slovenia, Greece, Malta, Cabo Verde, Italy, Botswana, Czech Republic, Portugal, Estonia, United States of America, Chile, Spain, Israel, France and Austria; *Hybrid regimes*: Mauritania, Nigeria, Benin, Côte d'Ivoire, Pakistan, Turkey, Gambia, Nepal, Uganda, Guatemala, Bolivia, Sierra Leone, Bosnia and Herzegovina, Morocco, Kenya, Honduras, Tanzania, Georgia, Liberia, Armenia, Senegal, Mexico, Ukraine, Fiji, Madagascar, Bhutan, Ecuador, El Salvador, Zambia, Malawi, Paraguay, Bangladesh and Tunisia; *Authoritarian*: Afghanistan, Myanmar, Democratic People's Republic of Korea, Democratic Republic of Congo, Central African Republic, Syrian Arab Republic, Turkmenistan, Chad, Lao People's Democratic Republic, Equatorial Guinea, Tajikistan, Islamic Republic of Iran, Libyan Arab Jamahiriya, Yemen, Eritrea, Saudi Arabia, Bolivarian Republic of Venezuela, Uzbekistan, Burundi, China, Guinea, Belarus, Sudan, Bahrain, Cameroon, Cuba, Azerbaijan, Nicaragua, Djibouti,

Guinea-Bissau, Congo, Togo, Cambodia, United Arab Emirates, Egypt, Viet Nam, Oman, Eswatini, Kazakhstan, Rwanda, Comoros, Niger, Russian Federation, Ethiopia, Angola, Gabon, Haiti, Mali, Jordan, Iraq, Mozambique, Kyrgyzstan, Qatar, Algeria, Burkina Faso, Lebanon and Kuwait.

References

Akbulut, B. (2019). The "State" of Degrowth: Economic Growth and the Making of State Hegemony in Turkey. *Environment and Planning E 2*(3), 513–527.

Amul, G.G.H., Tan, G.G.P. and van der Eijk, Y. (2021). A Systematic Review of Tobacco Industry Tactics in Southeast Asia: Lessons for Other Low-And Middle-Income Regions. *International Journal of Health Policy and Management 10*(6), 324–37.

ANZSOG (2021). Regulating Lobbying in Australia: Three Steps for Reform | ANZSOG. Retrieved from www.anzsog.edu.au/resource-library/news-media/opinion/regulating-lobbying-in-australia-three-steps-for-reform, 17 March 2022.

BBC News. (2020). Climate Change: US Formally Withdraws from Paris Agreement. Retrieved from www.bbc.com/news/science-environment-54797743, 24 April 2022.

Bell, D. and Building, A.R. (2014). What Is Liberalism? *Political Theory 42*(6), 682–715.

Bernhagen, P. (2013). When Do Politicians Listen to Lobbyists (and Who Benefits When They Do)? *European Journal of Political Research 52*(1), 20–43.

Blach-Ørsten, M., Willig, I. and Pedersen, L.H. (2017). PR, Lobbyism and Democracy. *Nordicom Review 38*(2), 19–31.

Boese, V.A. (2019). How (Not) to Measure Democracy. *International Area Studies Review 22*(2), 95–127.

Brand, U. (2016). "Transformation" as a New Critical Orthodoxy: The Strategic Use of the Term "Transformation" Does Not Prevent Multiple Crises. *GAIA 25*(1), 23–27.

Britannica. (2019). Checks and Balances | Definition, History, and Facts. *Encyclopedia Britannica.* Retrieved from www.britannica.com/topic/checks-and-balances, 25 April 2022.

Campbell, C. (2018). The Puzzle of Modern Consumerism. In *The Romantic Ethic and the Spirit of Modern Consumerism: New Extended Edition*, ed. Colin Campbell. Cham: Springer International Publishing, 77–105.

Canclini, N. (1995). *Consumidores y Ciudadanos Conflictos Multiculturales de La Globalización.* Ciudad de México: Grijalbo.

Chen, Y., Wang, Z., and Zhong, Z. (2019). CO_2 emissions, economic growth, renewable and non-renewable energy production and foreign trade in China. *Renewable Energy, 131*, 208–216

Choi, E., Heshmati, A. and Cho, Y. (2010). An Empirical Study of the Relationships Between CO_2 Emissions, Economic Growth and Openness. IZA Discussion Paper No.5304. Retrieved from https://papers.ssrn.com/sol3/papers.cfm?abstract_id=1708750, 24 April 2022.

Chomsky, N. (1999). *Profit Over People Neoliberalism and Global Order.* Seven Stories Press.

Chu, H, and Yang, J.Z. (2020). Risk or Efficacy? How Psychological Distance Influences Climate Change Engagement. *Risk Analysis 40*(4), 758–770.

Climate Action Tracker (2022). Climate Action Tracker. Retrieved from https://climateactiontracker.org/, 24 April 2022.

Cole, M.A., Rayner, A.J. and Bates, J.M. (2001). The Environmental Kuznets Curve: An Empirical Analysis. *Environment and Development Economics 2*(4), 401–416.

Cordoba, J. (2022). Costa Rica's new leader takes over with a blast at the past. Retrieved from https://abcnews.go.com/International/wireStory/costa-ricas-leader-takes-blast-past-84578131, 13 May 2022.

Cornell Law School (2022). Separation of powers | Wex | US Law | LII / Legal Information Institute. Retrieved from www.law.cornell.edu/wex/separation_of_powers_0, 25 April 2022.

Costanza, R., Hart, M., Kubiszewski, I., Posner, S., Talberth, J., Bell, S. and Morse, S. (2018). Lessons from the History of GDP in the Effort to Create Better Indicators of Prosperity, Well-Being, and Happiness. In *Routledge Handbook of Sustainability Indicators*, ed. S. Bell and S. Morse. London: Routledge, 117–123.

Dahl, R.A. (1971). *Polyarchy; Participation and Opposition*. New Haven, CT: Yale University Press.

Dalton, R.J., Sin, T-C. and Jou, W. (2007). Understanding Democracy: Data from Unlikely Places. *Journal of Democracy 18*(4), 142–156.

Dasgupta, S., Laplante, B., Wang, H. and Wheeler, D. (2002). Confronting the Environmental Kuznets Curve. *Journal of Economic Perspectives 16*(1). 147–168.

Dinda, S. (2004). Environmental Kuznets Curve Hypothesis: A Survey. *Ecological Economics 49*(4), 431–455.

The Economist Intelligence Group (2021). *Democracy Index 2020: In Sickness and in Health?* London: The Economist Intelligence Group.

Elff, M., and Ziaja, S. (2018). Method Factors in Democracy Indicators. *Politics and Governance, 6*(1), 92–104.

Feygina, I., Jost, J.T. and Goldsmith, R.E. (2009). System Justification, the Denial of Global Warming, and the Possibility of "System-Sanctioned Change." *Personality and Social Psychology Bulletin 36*(3), 326–338.

Friedlingstein, P., O'Sullivan, M., Jones, M.W., Andrew, R.M., Hauck, J., Olsen. A…… Zaehle, S. (2020). Global Carbon Budget 2020. *Earth System Science Data 12*(4), 3269–3340.

Gallup. (2021). Environment | Gallup Historical Trends. Retrieved from https://news.gallup.com/poll/1615/Environment.aspx, 11 April 2022.

Galston, W.A. (2018). The Populist Challenge to Liberal Democracy. *Journal of Democracy 29*(2), 5–19.

Gesami, B. (2021). Citizens United and Campaign Financing in the 2020 U.S Presidential Election. Retrieved from www.academia.edu/49521008/CITIZENS_UNITED_AND_CAMPAIGN_FINANCING_IN_THE_2020_U_S_PRESIDENTIAL_ELECTION, 15 March 2022.

Global Footprint Network (2021). *National Footprint and Biocapacity Accounts, 2021 Edition*. San Franscisco, CA: GFN.

Görg, C., Brand, U., Harbel, H., Hummel, D., Jahn, T. and Liehr, S. (2017). Challenges for Social-Ecological Transformations: Contributions from Social and Political Ecology. *Sustainability 9*(7), 21pp.

Harding, E. (2019). Spoilage and Squatting: A Lockean Argument. *Res Publica 26*(3), 299–317.

Hungrichhaußen, L. (2021). Successful Sustainability Policies—Analysing Energy and Transportation Policies of 12 Cities. Bachelor Dissertation. Breda University of Applied Sciences.

IPCC. (2017). FAQ Chapter 1—Global Warming of 1.5°C. *Special Report*. Retrieved from www.ipcc.ch/sr15/faq/faq-chapter-1/, 24, April 2022.

IPCC. (2022). Summary for Policymakers—Global Warming of 1.5 °C. *IPCC Summary*. Retrieved from www.ipcc.ch/sr15/chapter/spm/, 24 April 2022.

Iyer, R. and Muncy, J.A. (2016). Attitude toward Consumption and Subjective Well-Being. *Journal of Consumer Affairs 50*(1), 48–67.

Jasanoff, S. (2010). A New Climate for Society. *Culture and Society 27*, 233–253.

Jones, C., Hine, D.W. and Marks, A.D.G. (2017). The Future Is Now: Reducing Psychological Distance to Increase Public Engagement with Climate Change. *Risk Analysis 37*(2), 331–341.

Karsch, N.M. (2019). Examining the Validity of the Environmental Kuznets Curve. *Consilience 21*(1), 32–50.

Kasman, A. and Duman, Y.S. (2015). CO_2 Emissions, Economic Growth, Energy Consumption, Trade and Urbanization in New EU Member and Candidate Countries: A Panel Data Analysis. *Economic Modelling 44*, 97–103.

Kuznets, S. (1955). Economic Growth and Income Inequality. *The American Economic Review 45*(1), 1–28.

Labour Party (2022). Rt Hon Trevor Mallard—List MP Based in Hutt South—NZ Labour Party. Retrieved from www.labour.org.nz/trevormallard, 17 March 2022.

Levitt, J. (2011). Confronting the Impact of Citizens United. *Yale Law and Policy Review 29*(1), 217–234.

Maglio, S.J., Trope, Y. and Liberman, N. (2013). Distance from a Distance: Psychological Distance Reduces Sensitivity to Any Further Psychological Distance. *Journal of Experimental Psychology 142*(3), 644–657.

de Mesquita, B. and Smith, A. (2011). *The Dictator's Handbook—Why Bad Behavior Is Almost Always Good Politics*. New York: PublicAffairs.

Moloney, K. and McGrath, C. (2019). *Rethinking Public Relations: Persuasion, Democracy and Society*. Abingdon, NY: Routledge.

Mudge, S.L. (2008). What Is Neo-Liberalism? *Socio-Economic Review 6*(4), 703–731.

Nahman, A. and Antrobus, G. (2005). The Environmental Kuznets Curve: A Literature Survey. *South African Journal of Economics 73*(1), 105–120.

New Zealand Government (2022). Lobbying and the Government. Retrieved from www.beehive.govt.nz/speech/lobbying-and-government, 17 March 2022.

New Zealand Parliament (2022). Lobbying Disclosure Bill—New Zealand Parliament. Retrieved from www.parliament.nz/en/pb/bills-and-laws/bills-proposed-laws/document/00DBHOH_BILL11278_1/lobbying-disclosure-bill, 17 March 2022.

O'Brien, K. 2011. Global Environmental Change II: From Adaptation to Deliberate Transformation. *Progress in Human Geography 36*(5), 667–676.

Our World in Data (2015). GDP per Head vs Share of Industry in Employment, 1800 to 2015. Retrieved from https://ourworldindata.org/grapher/gdp-vs-manufacturing-employment?country=USA~GBR~DEU~FRA~CHN~IND~IDN~TWN, 24 April 2022.

Our World in Data (2019a). CO_2 Emissions. Retrieved from https://ourworldindata.org/co2-emissions, 24 April 2022.

Our World in Data (2019b). Consumption-Based CO_2 Emissions per Capita vs GDP per Capita. Retrieved from https://ourworldindata.org/grapher/consumption-co2-per-capita-vs-gdppc, 24 April 2022.

Our World in Data (2019c). Smoking Death Rate in 1990 vs. 2019. Retrieved from https://ourworldindata.org/grapher/smoking-death-rate-1990-2017, 17 March 2022.

Our World in Data (2022). Economic Growth. Retrieved from https://ourworldindata.org/economic-growth?ref=hackernoon.com, 22 April 2022.

Pearce, W., Grundmann, R., Hulme, M., Sujatha, R., Hadley Kershaw, E. and Tsouvalis, J. (2017). Beyond Counting Climate Consensus. *Environmental Communication 11*(6), 723–730.

Phelan, S., and Dawes, S. (2018). Liberalism and Neoliberalism. In *Oxford Research Encyclopedia of Communication*. Oxford University Press.

Pichler, M., Brand, U. and Görg, C. (2018). The Double Materiality of Democracy in Capitalist Societies: Challenges for Social-Ecological Transformations. *Environmental Politics 29*(2), 193–213.

Pichler, M., Schaffartzik, A., Haberl, H. and Görg, C. (2017). Drivers of Society-Nature Relations in the Anthropocene and Their Implications for Sustainability Transformations. *Current Opinion in Environmental Sustainability 26–27*, 32–36.

Piketty, T. (2017). *Capital in the Twenty-First Century*. Cambridge, MA: Harvard University Press.

Russell, D., Schin, D. and Jou, W. (2007). Understanding Democracy: Data from Unlikely Places. *Journal of Democracy 18*(4), 142–156.

Samet, D. and Schmeidler, D. (2003). Between Liberalism and Democracy. *Journal of Economic Theory 110*(2), 213–233.

Schimitter, P. and Karl, T.L. (1991). What Democracy Is…and Is Not. *Journal of Democracy 2*(3), 75–88.

Schumpeter, J. (1943). *Capitalism, Socialism and Democracy*. London: Routledge.

Sheppard, S.R.J., Shaw, A., Flanders, D., Burch, S., Wiek, A., Carmichael, J., Robinson, J. and Cohen, S. (2011). Future Visioning of Local Climate Change: A Framework for Community Engagement and Planning with Scenarios and Visualisation. *Futures 43*(4), 400–412.

Shwed, U. and Bearman, P.S. (2010). The Temporal Structure of Scientific Consensus Formation. *American Sociological Review 75*(6), 817–840. www.ncbi.nlm.nih.gov/pubmed/21886269 March 17, 2022.

SOS Children's Villages (n.d). *Poverty in Africa—The Indicators*. Retrieved from www.sos-usa.org/about-us/where-we-work/africa/poverty-in-africa#:~:text=Africa%20is%20considered%20the%20poorest,society%2C%20their%20children%20and%20women, 10 May 2022.

Spence, A., Poortinga, W. and Pidgeon, N. (2012). The Psychological Distance of Climate Change. *Risk Analysis 32*(6), 957–972.

Stephens, P.H.G. (2001) Debate—Green Liberalisms: Nature, Agency and the Good. *Environmental Politics 10*(3), 1–22. https://doi.org/10.1080/714000559.

Stern, D.I. (2004). The Rise and Fall of the Environmental Kuznets Curve. *World Development 32*(8), 1419–1439.

Subedi, D.B. and Scott, A. (2021). Populism, Authoritarianism, and Charismatic-Plebiscitary Leadership in Contemporary Asia: A Comparative Perspective from India and Myanmar. *Contemporary Politics 27*(5), 487–507.

Teachout, Z. (2010). Facts in Exile: Corruption and Abstraction in *Citizens United v. Federal Election Commission*. *Loyola University Chicago Law Journal 42*(2), 295.

Thorsen, D.E. and Lie, A. (2007). What Is Neoliberalism. In *Neoliberalism Ideas and Political Reality*, ed. Dag Harald Claes et al. Oslo: Universitetsforlaget, 33–48.

Trope, Y. and Liberman, N. (2010). Construal-Level Theory of Psychological Distance. *Psychological Review 117*(2), 440–463.

United Nations (2022a). Human Development Index. Retrieved from www.who.int/data/nutrition/nlis/info/human-development-index, 15 March 2022.

United Nations (2022b). WGII Summary for Policymakers Headline Statements. Retrieved from www.ipcc.ch/report/ar6/wg3/resources/spm-headline-statements, 14 April 2022.

UNDP (United Nations Development Programme) (n.d.). Human Development Reports. Retrieved from https://hdr.undp.org/en/content/human-development-report-2020-readers-guide#:~:text=Human%20development%20classification&text=The%20cutoff%2Dpoints%20are%20HDI,for%20very%20high%20human%20development, 13 May 2022.

V-Dem (2022). Varieties of Democracy. Retrieved from www.v-dem.net/, 15 May 2022.
Verly, H. and Ruiz, J. (2022). Government Lobbying in Argentina—Lexology. Retrieved from www.lexology.com/library/detail.aspx?g=1c697958-02c1-4a42-af18-b30458ead427, 17 March 2022.
Vestheim, G. (2012). Cultural policy and democracy: an introduction. *International Journal of Cultural Policy*, *18*(5), 493–504.
Watts, C., Burton, S. and Freeman, B. (2020). Creating a Market for IQOS: Analysis of Philip Morris' Strategy to Introduce Heated Tobacco Products to the Australian Consumer Market. *Tobacco Control*. Retrieved from https://tobaccocontrol.bmj.com/content/early/2020/11/07/tobaccocontrol-2020-056057, 17 March 2022.
Wilson Center (2022). Lobbying and Corruption in Democratic Brazil. Retrieved from www.wilsoncenter.org/event/lobbying-and-corruption-democratic-brazil, 17 March 2022.

10

CAPITALISM

10.1 Introduction

Back in the early 1990s during the 1992 US election campaign the phrase, "It's the economy, stupid" was devised by Bill Clinton's strategist James Carville (Foroohar, 2022). If the economy is at the base of everything, how does it impact on the environment and on cities? In 1905 the German sociologist Max Weber wrote in his influential book *The Protestant Ethic and the Spirit of Capitalism*, "the modern economic order…is now bound to the technical and economic conditions of machine production which today determine the lives of all…with irresistible force. Perhaps it will so determine them until the last ton of fossilized coal is burnt" (Weber, 1905:181).

Weber thought that capitalism would endure until there was no more coal left to fuel it. Now we are more concerned about the environmental impact of burning that "last ton of fossilized coal". In a book published in 1999, the Nobel prize-winning economist Amartya Sen wrote: "The solutions to these problems—inequality…and…the environment…will almost certainly call for institutions that take us beyond the capitalist market economy" (Sen, 1999:267). More recently an article published by the World Economic Forum concludes "capitalism as an economic system simply will not safeguard our planet" (Movahed, 2016). Both these statements suggest that it is specifically the capitalist economy that is the problem for the environment.

10.2 What is capitalism?

Is capitalism really the problem or is it something else? We often refer to capitalism, but it is often not defined. What is it that makes you a capitalist? The website capitalism.org might be considered the logical place to look for a definition

DOI: 10.4324/9781003045847-12

of what capitalism is all about. According to this website, "Capitalism is the social system based on the individual's right to life, liberty, property, and the pursuit of happiness" (Capitalism.org, 2021). This definition is very similar to the part of the United States' Declaration of Independence about "certain unalienable Rights", which states: "We hold these truths to be self-evident, that all men are created equal, that they are endowed by their Creator with certain unalienable Rights, that among these are Life, Liberty and the pursuit of Happiness" (National Archives, 2021). In this website's definition, the Declaration of Independence is used to imply that capitalism is fundamental to the United States.

From the opposing socialist viewpoint, capitalism is the problem:

> It is not the working class and the poor, nor the middle classes either, who are responsible for the possibility of runaway climate change. It has been under the capitalists' stewardship of society that the planet has been brought to the perilous position it is in today
>
> *Socialism Today, 2021*

The website for capitalism perhaps represents an extreme view of what capitalism means as a philosophy. It seems all very well to have these rights, but in a capitalist world you can have them only if you have the money to buy them. Your liberty and happiness and even your life may well be constrained by poverty, which will certainly constrain your right to acquire property.

Writing about Weber's work, The Oxford University historian Peter Ghosh wrote that capitalism had become an ethic.

> [Is is] the code of values by which people lived in the 20th-century West, and now live, we may add, in much of the 21st-century globe…a situation where the "highest good" was the making of money and ever more money, without any limit…is what is most readily recognisable as the "spirit" of capitalism.
>
> *Ghosh, 2018*

Capitalism is basically a system for organising the economy, so it could be useful to see how it is defined in economic terms rather than as a philosophy. A simple definition is that capitalism is "an economic system in which investment in and ownership of the means of production, distribution, and exchange of wealth is made and maintained chiefly by private individuals or corporations, especially as contrasted to cooperatively or state-owned means of wealth" (Dictionary.com, 2022a). The focus on private individuals reflects the earlier definition above, but what are corporations? The same source says that a corporation is "an association of individuals, created by law or under authority of law, having a continuous existence independent of the existences of its members, and powers and liabilities distinct from those of its members" (Dictionary.com, 2022b). When a group of individuals come together to form a corporation, they create something that exists outside themselves that can continue beyond their individual lives; Colonel Sanders is dead, but KFC lives on (Biography.com Editors, 2014).

A more detailed analysis, provided by the International Monetary Fund (IMF) (Jahan and Mahmud, 2015), cites the six "pillars of capitalism" with brief definitions of how each one operates. These definitions are worth examining in more detail because they provide insights into some of capitalism's problems.

The first pillar is "private property, which allows people to own tangible assets such as land and houses and intangible assets such as stocks and bonds". This is fine, provided you can afford to own any of these tangible or intangible assets.

The second pillar is

> self-interest, through which people act in pursuit of their own good, without regard for sociopolitical pressure. Nonetheless, these uncoordinated individuals end up benefiting society as if, in the words of Smith's 1776 Wealth of Nations, they were guided by an invisible hand.

Given the increasing problems of climate change, pollution and inequality, it would seem that these "uncoordinated individuals" are not ending up "benefiting society".

Next, we have "competition, [which] through firms' freedom to enter and exit markets, maximizes social welfare, that is, the joint welfare of both producers and consumers". Given the relative power imbalance between large producer organisations on the one hand and individuals on the other, is it possible to argue that "the joint welfare of both producers and consumers" is taking place?

The fourth pillar is "a market mechanism that determines prices in a decentralized manner through interactions between buyers and sellers—prices, in return, allocate resources, which naturally seek the highest reward, not only for goods and services but for wages as well". This pillar also implies more of a balance between buyers and sellers and wage earners than seems to exist in practice.

The next pillar is "freedom to choose with respect to consumption, production, and investment—dissatisfied customers can buy different products, investors can pursue more lucrative ventures, workers can leave their jobs for better pay." Setting aside the idea that workers can really leave their job in order to obtain better pay, choice may be somewhat illusory if you cannot afford anything but the cheapest.

The final pillar of capitalism is the "limited role of government, to protect the rights of private citizens and maintain an orderly environment that facilitates proper functioning of markets".

10.3 Capitalism, growth and the environment

There appears to be nothing in these six pillars of capitalism that could be said to threaten the environment, but neither is there anything to benefit it. From the perspective of the environment—and it should be acknowledged that at the time of writing this chapter the Doomsday Clock of the *Bulletin of the Atomic Scientists* "remains the closest it has ever been to civilization-ending apocalypse" in large part due to the lack of response to climate change (Mecklin, 2022)—the six pillars of capitalism appear to have very little to offer. They mention welfare and benefiting society, but the focus is largely on ownership, markets and rewards, which are assumed to bring these benefits.

The article provided by the IMF states at the outset "Free markets may not be perfect but they are probably the best way to organize an economy". Its authors state that "the essential feature of capitalism is the motive to make a profit" (Jahan and Mahmud, 2015). Profit is "the monetary surplus left to a producer or employer after deducting wages, rent, cost of raw materials, etc" (Dictionary.com, 2022c). Why does the *Bulletin of the Atomic Scientists* seem so gloomy? Perhaps the root of the problem is economic growth, "the motive to make a profit", to produce a "monetary surplus". If we have a rising global population all seeking a monetary surplus, we will have rising material consumption and rising environmental damage. This argument sounds glibly convincing, but is it true? An article on economic growth by the Oxford University "Our World in Data" project (Roser, 2013) defines growth as "an increase in the quantity and quality of the economic goods and services that a society produces and consumes". Roser goes on to explain:

> What economic growth makes possible is that everyone can become better off, even when the number of people that need to be served by the economy increases. An almost 3-fold increase of the population multiplied by a 4.4-fold increase in average prosperity means that the global economy has grown 13-fold since 1950.

Roser goes on to discuss the need to achieve economic growth while reducing environmental impact, maintaining that better health care, mental health and education, finding alternatives to eating meat and developing better solar energy technology could all be ways to have growth with reduced environmental impact. He concludes optimistically: "It is not the case anymore that one person's or one country's gain is automatically another one's loss" (Roser, 2013).

This view of economists that we can all have more without any consequences fits uncomfortably with another way of viewing the world, this time from an ecological rather than an economic perspective. The ecological footprint (EF; see Chapter 2) is a measure of how much of the environment we need to produce all the goods and services that make up our way of living. Since 1972, the date of the United Nations' Stockholm Conference on the Environment, "the first world conference to make the environment a major issue" (United Nations, n.d), the world population has risen from around four billion to nearly eight billion (Worldometer, 2022a) while the area of land has remained the same. Unlike the economy, the environment is fixed, there is "only one Earth" (Ward and Dubos, 1972). Over the last 50 years, the same environment has had to meet the needs of twice as many people, making the idea that we can all have more seem rather unconvincing.

The gap between humanity's demands and the environment available to meet them is termed the "ecological overshoot." According to the Global Footprint Network (2022a), the organisation set up to further the use of EF accounting:

> Today humanity uses the equivalent of 1.75 Earths to provide the resources we use and absorb our waste…it now takes the Earth one year and eight months to regenerate what we use in a year. Humanity uses more ecological resources and services than nature can regenerate through overfishing, overharvesting forests, and emitting more carbon dioxide into the atmosphere than forests can sequester.

This is not a situation that can continue indefinitely. Commenting on the United Nations Conference on the Environment, COP 26, held in Glasgow in 2021, Bill Rees, the co-originator of the ecological footprint, wrote "we must acknowledge that the human ecological footprint, including the overshoot portion, is the product of average material consumption × population". He concluded that being able to live within the limits of our environment would mean a 40% global reduction in energy and material consumption per person, which would require an 80% reduction in high-income countries; it would need more equitable sharing and "a smooth, socially just descent" to a population of one to two billion. "The overall goal needs to be a smaller, steady-state global economy/society of fewer people living more equitably and securely within the biophysical means of nature" (Rees, 2021).

Rees' steady-state society of one or two billion people is a long way indeed from the idea that all can have more in a world of eight billion. Even the World Bank has found that increasing wealth has problems. "Global wealth has grown overall—but at the expense of future prosperity and by exacerbating inequalities" (World Bank, 2021b).

10.4 The role of growth

It may not even be capitalism that is to blame, a study of environmental damage in communist East and capitalist West Germany concluded that the damage was the result of economic growth.

> In the 1950s and 1960s, both East and West Germany had extraordinarily high rates of growth, and the environment suffered badly in both parts of the nation. Over the next two decades, however, growth rates under capitalism fell by approximately one-half, and the condition of the environment in the West improved somewhat. The communists, on the other hand, struggled to keep the growth rate high, sacrificing the environment in the process.
>
> *Dominick, 1998:324*

In this analysis, it is not the economic system but economic growth, regardless of the system, that creates environmental damage. It is quite hard to imagine how economic growth would not result in environmental damage. Growth means having more and having more will require the use of more resources. This is the same whether the growth is in population or in consumption. Is this what has happened in practice? Over the period from 1958 to 2022, the world population has increased from nearly 3 billion (Worldometer, 2022a) to nearly 8 billion (Worldometer, 2022b), an increase of 167%. Even if no one's material standard of living had increased in those past 60-odd years, 8 billion people are going to need more than twice as much food, and twice as much land on which to grow it, as 3 billion people.

One clear measure of environmental damage is the level of carbon dioxide in the atmosphere, as this is the gas that is warming the climate. Since the start of systematic atmospheric carbon dioxide records in 1958, the carbon dioxide in the atmosphere has risen from 315 ppm (parts per million) to 418 ppm (US Department of Commerce, 2022), an increase of about 33%. The population is more than double what it was in 1958, but carbon dioxide emissions have increased by only a third. This happens because the rapid population expansion has been largely in poorer countries where people use much less energy (see also Section 11.3). Generally, the poor have been getting a little richer, but the rich have got a lot richer (Vale and Vale, 2009:22).

Growth in population is one aspect of growth, the other is economic growth. World GDP has increased from $13.89 trillion in 1959 to $108.12 trillion in 2015 (both values in 2011 dollars to allow comparison, the dates are those available, they do not quite match the dates of 1958 to 2022 used above) (Our World in Data, 2022). Basically, while the population doubled, GDP increased nearly eight times over the same period. This looks like a great success for the world's economic system—there are a lot more of us, we all appear to be much richer and emissions have only increased a bit, so what is the problem?

Perhaps that kind of thinking is at the root of the problem. From the perspective of the ecological footprint (see Chapter 2), the situation looks very different. EF is a measure of all activity, not just of carbon emissions and it is based on the land available to meet humanity's needs. In 1961, the earliest year for which EFs have been calculated, the global EF was 2.27 gha and the Earth's biocapacity, the land available to provide all our needs, was 3.11 gha per person. In 1961 humanity still had more than enough land to support its needs. The land needed to absorb carbon emissions was 1.0 gha per person. In 2017, the last year for which data are published, the global EF had risen to 2.77 gha, the Earth's biocapacity had fallen to 1.6 gha per person and the land needed to absorb carbon had risen to 1.69 gha per person (Global Footprint Network, 2022b). In 1961 the whole world was living within its capacity and the land needed to absorb humanity's carbon emissions was 44% of the total footprint. By 2017 the EF of the average person worldwide was a lot larger than the biocapacity available to support them and carbon land formed 61% of the total footprint.

Whereas, as seen earlier, from an economist's perspective, "it is not the case anymore that one person's or one country's gain is automatically another one's loss"

(Roser, 2013), the perspective of ecological footprint, being based on available land area, means if I have more, you, by definition, will end up having less.

The EFs of citizens of the 19 different countries, which are the locations of the cities in this book, are compared in Table 10.1, where the countries are ranked by GDP. Bold font shows countries with footprints below their national biocapacity and italics those with footprints below global biocapacity (1.6 gha, 2017).

Table 10.1 suggests very clearly that environmental damage, as measured by the size of the ecological footprint, is linked to wealth. The poorer countries in general have the lowest footprints, although it is encouraging that this is not a hard-and-fast rule; the Netherlands have a footprint only 70% of that of Australia combined with a higher GDP, similarly New Zealand's footprint is only half that of the UAE in spite of having a larger GDP. The greater problem is that even these relatively well-performing wealthy countries have footprints that are far too high to be sustainable, the only three countries with footprints that are less than the world's available biocapacity of 1.6 gha are the very poorest.

Table 10.2 shows that back in 1961, when the global average footprint was 2.3 gha, the biocapacity was 3.1 gha (Global Footprint Network, 2022b) and the

TABLE 10.1 Ecological footprint and biocapacity per head compared with National GDP (countries listed in order of GDP)

Country ranked by GDP	*Ecological footprint gha per head*[1]	*Biocapacity (gha per head)*[1]	*GDP per head 2020 (US$)*[2]
United States	8.1	3.5	63,413
Denmark	6.9	4.2	61,063
Singapore	5.9	0.06	59,798
Netherlands	5.0	0.8	52,397
Australia	**7.1**	**12.5**	**51,693**
Germany	4.7	1.6	46,208
New Zealand	**4.4**	**9.2**	**41,442**
UAE	8.9	0.5	36,285
China	3.7	0.9	10,435
Malaysia	3.9	2.3	10,412
Russian Federation	**5.4**	**6.9**	**10,127**
Cuba	1.8	0.8	9,478
Argentina	**3.3**	**6.6**	**8,579**
Brazil	**2.8**	**8.6**	**6,797**
Indonesia	1.7	1.2	3,870
Egypt	1.8	0.4	3,569
India	*1.2*	*0.4*	*1,928*
Kenya	*1.0*	*0.5*	*1,879*
North Korea	*0.8*	*0.6*	*N/A*
WORLD	2.8	1.6	10,919

1 Global Footprint Network (2022c).
2 World Bank Group (2022).

TABLE 10.2 Ecological footprint and biocapacity per head in the 19 countries in 1961 and 2017

Country	*Ecological footprint (gha per head),[1] 1961*	*Ecological footprint (gha per head),[1] 2017*	*Biocapacity,[1] 1961*	*Biocapacity,[1] 2017*	*GDP per head (US$)[2]*
United States	8.0	8.1	5.0	3.5	
Denmark	5.8	6.9	4.5	4.2	
Singapore	N/A	5.9	N/A	0.06	
Netherlands	3.5	5.0	1.2	0.8	
Australia	7.5	7.1	29.6	12.5	
Germany	4.3	4.7	1.3	1.6	
New Zealand	4.8	4.4	18.6	9.2	
UAE	6.6 (1980)	8.9	4.6 (1980)	0.5	
China	0.9	3.7	1.0	0.9	
Malaysia	1.4	3.9	6.1	2.3	
Russian Federation	6.7 (1992)	5.4	6.7 (1992)	6.9	
Cuba	1.5	1.8	1.0	0.8	
Argentina	3.8	3.3	10.4	6.6	
Brazil	2.4	2.8	22.9	8.6	
Indonesia	1.5	1.7	2.7	1.2	
Egypt	0.8	1.8	0.4	0.4	
India	0.6	1.2	0.5	0.4	
Kenya	2.2	1.0	1.8	0.5	
North Korea	1.1	0.8	1.3	0.6	
World 1961	2.3		3.1		$470
World 2017	2.8		1.6		$10,919

1 Global Footprint Network (2022c).
2 World Bank Group (2022).

global GDP was only $470 per head (World Bank Group, 2022), there were 8 countries out of the 19 we are studying living within the world's biocapacity. Bold shows countries with footprints below their national biocapacity in 1961 and bold italic countries with footprints below global biocapacity (2.3 gha in 1961).

Over the 56 years from 1961 to 2017, every person in the world has become, on average, over 20 times wealthier. At the same time, the area of land available to support them, the biocapacity, has halved. We are reducing poverty but at the cost of damaging the whole planet, in ways which may make it far less habitable.

10.5 Capitalism and cities

As we saw in Chapter 1, worldwide, 55% of people were urban dwellers in 2018, a figure predicted to rise to 68% in 2025 (UN, 2018), while the rural population (who grow the food for the people in the cities) dropped from just over 66% of the total in 1960 to just under 45% in 2018 (The World Bank Group, 2019). Partly because they rely on bringing in resources from elsewhere, cities often have higher EFs than their national average (see Chapter 1), at a time when humanity is already using up the resources of the planet at a faster rate than they are being replenished (GFN, 2021a).

Cities are the places where wealth is created (Jacobs, 1984), so it is not surprising that cities often have higher footprints than the nations in which they are located. In a detailed study of 19 cities located around the Mediterranean, Baabou et al. (2017) found that most of these cities had a larger footprint than their national footprint, with the largest footprints being in high-income countries. They found the three principal components of the urban footprint were food, personal transport and goods, with food having greater impact in lower-income countries. In high-income countries, transport was a greater part of the total footprint than in lower-income ones.

The same authors provided results from a wide range of city footprint studies from around the world (Baabou et al., 2017). These results are shown here in Table 10.3, which compares city footprints with national footprints for the same date, along with available global biocapacity for the same date. In this dataset, there are 30 cities out of the 44 in Table 10.3 that have footprints lower than the countries in which they are situated. This is encouraging. What is more worrying is that most of the cities have footprints that are three or four times larger than the world's biocapacity. Given that global biocapacity represents the maximum amount of land available to support human activities, only 3 cities out of the 44 in Table 10.3 come even close to being able to be regarded as being sustainable. Because the studies quoted by Baabou et al. are all 20–25 years old (see Table 10.3) it is highly likely that now even fewer cities could be considered sustainable. There may also be discrepancies in the way the EFs have been calculated. In Table 10.3, italic font shows cities below or close to global biocapacity; bold font shows cities with lower footprints than the countries in which they are located. City data are from Baabou et al. (2017: Table 1:96) and world and national footprint data are from Global Footprint Network (2022c).

TABLE 10.3 Ecological footprints of a range of cities from many studies, compared with national and global EFs for the same date as each study

Country	*Date of study*	*City*	*National footprint at the same date (gha)*	*City footprint (gha)*	*World biocapacity at the same date (gha)*
Australia	**2001**	**Sydney**	**8.1**	**5.92**	**1.8**
Brazil	**2009**	**Curitiba**	**2.7**	**2.60**	**1.7**
Canada	2001	Calgary	8.7	9.86	1.8
	2007	Calgary	8.7	9.5–9.9	1.7
	2001	Edmonton	8.7	9.45	1.8
	2008	Edmonton	8.4	8.56	1.7
	2001	**Québec City**	**8.7**	**6.89**	**1.8**
	2001	**Toronto**	**8.7**	**7.36**	**1.8**
	2006	**Vancouver**	**8.9**	**4.70**	**1.7**
	2001	**Vancouver**	**8.7**	**7.71**	**1.8**

TABLE 10.3 Cont.

Country	*Date of study*	*City*	*National foot-print at the same date (gha)*	*City foot-print (gha)*	*World biocapacity at the same date (gha)*
Chile	**1998**	**Santiago de Chile**	**3.6**	**2.60**	**1.9**
China	**2009**	**Chongqing**	**3.1**	**2.20**	**1.7**
	2008	Hong Kong	2.9	4.30	1.7
	2009	Shanghai	3.1	3.80	1.7
	2009	***Shenyang***	***3.1***	***1.80***	***1.7***
	2009	**Tianjin**	**3.1**	**2.70**	**1.7**
Ecuador	2006	Quito	1.8	2.40	1.7
Iran	***2007***	***Isfahan***	***3.2***	***1.22***	***1.7***
	2005	Tehran	3.0	3.79	1.7
Israel	**2007**	**Beer-Sheva**	**5.7**	**3.98**	**1.7**
	2002	**Ra'anana***	**5.7**	**4.00ha**	**1.8**
Italy	**2002**	**Piacenza**	**5.5**	**3.79**	**1.8**
	1999	Siena **	5.5	5.80	1.9
Japan	2009	Kawasaki	4.6	5.10	1.7
Norway	2000	Oslo	6.3	7.76	1.8
Philippines	*2009*	*Manila*	*1.2*	*1.82*	*1.7*
Spain	**1996**	**Barcelona***	**4.8**	**3.23ha**	**1.9**
United Kingdom	**2007**	**Birmingham**	**6.0**	**5.22**	**1.7**
	2007	**Bradford**	**6.0**	**5.21**	**1.7**
	2007	**Bristol**	**6.0**	**5.22**	**1.7**
	2007	**Cardiff**	**6.0**	**5.20**	**1.7**
	2001	**Cardiff**	**6.0**	**5.50**	**1.8**
	2007	**Edinburgh**	**6.0**	**5.76**	**1.7**
	2007	**Glasgow**	**6.0**	**5.21**	**1.7**
	2003	**Greater Nottingham**	**5.9**	**5.27**	**1.8**
	2007	**Liverpool**	**6.0**	**5.25**	**1.7**
	2007	**London**	**6.0**	**5.48**	**1.7**
	2007	**Manchester**	**6.0**	**5.36**	**1.7**
	2007	**Newport**	**6.0**	**5.01**	**1.7**
	2007	**Nottingham**	**6.0**	**5.26**	**1.7**
	2007	**Plymouth**	**6.0**	**5.01**	**1.7**
	2007	Winchester	6.0	6.52	1.7
	2000	York*	6.0	6.98ha	1.8
United States	**2007**	**San Francisco**	**9.9**	**7.10**	**1.7**

* Footprint results for the cities of Barcelona, Ra'anana and York are expressed in hectares rather than global hectares.

** The study of Siena also includes the Footprint of the other 35 municipalities belonging to the Provincial administration.

10.5.1 Wealth and cities

Wealth is created in cities as Jane Jacobs showed in 1984, and it is the growth in wealth that leads to the growth in environmental damage. As shown in Table 10.1, generally the richer the country the higher its ecological footprint. The political system apparently has less effect on environmental damage than economic growth, as shown by Dominick's comparison of West and East Germany (1998). From the end of the Second World War, until the mid-1970s, the USSR and the United

States had similar rates of economic growth, although the GDP of the United States was always greater than that of the Soviet Union (Harrison, 2017:Figure 2).

Wealth is still being created in cities, as stated in a report by the global investment management company PGIM (2017:20): "The mega-trend of urban expansion opens up important new investment opportunities for institutional investors...Investors should consider capitalizing on the opportunities now arising from this unprecedented population shift." This report suggests four major themes for urban investors which are broken down into ten categories as follows:

Urban infrastructure

- Wired cities
- Mega-city connectivity
- Anti-pollution
- Replacing and improving old assets

Real estate

- Urban revival
- Economical housing options
- Retail outlets and logistics support

Consumer goods and services

- New consumer class
- Global luxury

Agriculture supply chain

- Industrialised agriculture

While the report's detailed discussion of these themes mentions renewable energy, affordable housing and railways, much of the focus for investment is on airports and roads, shopping malls and consumer goods. Its disregard for environmental impact is shown most clearly in its discussion of investment in agriculture: "Given the explosive growth of middle-class consumers around the world, the demand for food—particularly meat and dairy—will escalate sharply over the coming decades" (PGIM, 2017:17). The Food and Agriculture Organisation of the United Nations (Gerber et al., 2013:15) found that livestock farming represented 14.5% of all human greenhouse gas emissions in 2004. In spite of this, PGIM promotes increased demand for the products of livestock farming as a good investment opportunity. Capitalism has become the guiding ethic of modern times with the sole aim of "the making of money and ever more money, without any limit"

(Ghosh, 2018). The result is that what matters is making money, while all other considerations are set aside.

Given that cities are the creators of wealth in an economy and given that the capitalist ethic considers only money, it is no surprise that there are so few cities in Table 10.3 that fall within the limit of global biocapacity. Unfortunately, it looks as though capitalism may be the death of us.

References

Baabou, W., Grunewald, N., Ouellet-Plamondon, C., Gressot, M. and Galli, A. (2017). The Ecological Footprint of Mediterranean Cities: Awareness Creation and Policy Implications. *Environmental Science & Policy 69*, 94–104.

Biography.com Editors (2014). Colonel Harland Sanders Biography. Retrieved from www.biography.com/business-figure/colonel-harland-sanders, 27 October 2021.

Capitalism.org (2021). Capitalism Tour. Retrieved from www.capitalism.org/, 28 January 2022.

Dictionary.com (2022a). Capitalism. Retrieved from www.dictionary.com/browse/capitalism, 28 January 2022.

Dictionary.com (2022b). Corporation. Retrieved from www.dictionary.com/browse/corporation, 28 January 2022.

Dictionary.com (2022c). Profit. Retrieved from www.dictionary.com/browse/profit, 18 February 2022.

Dominick, R. (1998). Capitalism, Communism, and Environmental Protection: Lessons from the German Experience. *Environmental History 3*(3), 311–32.

Foroohar, R. (2022). It's Not Just the Economy, Stupid. *Financial Times*, February 2021. Retrieved from www.ft.com/content/f6d53cd7-e384-457e-8b6d-fc7a815dda1c, 7 March 2022.

Gerber, P.J., Steinfeld, H., Henderson, B., Mottet, A., Opio, C., Dijkman, J., Falcucci, A. and Tempio, G. (2013). *Tackling Climate Change Through Livestock – A Global Assessment of Emissions and Mitigation Opportunities*. Rome: Food and Agriculture Organization of the United Nations (FAO).

Ghosh, P. (2018). What Did Max Weber Mean by the "Spirit" of Capitalism? *Aeon* 12 June. Retrieved from https://aeon.co/ideas/what-did-max-weber-mean-by-the-spirit-of-capitalism, 8 March 2022.

Global Footprint Network (2022a). World Footprint. Retrieved from www.footprintnetwork.org/our-work/ecological-footprint/, 7 February 2022.

Global Footprint Network (2022b). Compare Countries. Retrieved from https://data.footprintnetwork.org/?_ga=2.170585888.1043366151.1645150675-1048960784.1643677418#/compareCountries?type=EFCpc&cn=all&yr=1961, 18 February 2022.

Global Footprint Network (2022c). Interactive Global Map. Retrieved from https://data.footprintnetwork.org/?_ga=2.265119951.1702695995.1645917177-1048960784.1643677418#/, 3 March 2022.

Harrison, M. (2017). The Soviet Economy, 1917–1991: Its Life and Afterlife. Retrieved from https://voxeu.org/article/soviet-economy-1917-1991-its-life-and-afterlife, 28 February 2022.

Jahan, S. and Mahmud, A. (2015). What Is Capitalism? *Finance & Development 52*(2). Retrieved from www.imf.org/external/pubs/ft/fandd/2015/06/basics.htm, 28 January 2022.

Jacobs, J. (1984). *Cities and the Wealth of Nations*. New York: Random House.

Mecklin, J. (ed). (2022). At Doom's Doorstep: It is 100 Seconds to Midnight: 2022 Doomsday Clock Statement. Science and Security Board *Bulletin of the Atomic Scientists*. Retrieved from https://thebulletin.org/doomsday-clock/current-time/, 1 February 2022.

Movahed, M. (2016). Does Capitalism Have to Be Bad for the Environment? Geneva, World Economic Forum. Retrieved from www.weforum.org/agenda/2016/02/does-capitalism-have-to-be-bad-for-the-environment/, 8 February 2022.

National Archives (2021). Declaration of Independence: A Transcription. Retrieved from www.archives.gov/founding-docs/declaration-transcript, 28 January 2022.

Our World in Data (2022). World GDP Over the Last Two Millennia. Retrieved from https://ourworldindata.org/economic-growth, 18 February 2022.

PGIM (2017). *The Wealth of Cities: The Investment Implications of Urban Expansion*. Newark NJ: PGIM Inc.

Rees, W. (2021). Bill Rees: A Note on Climate Change and Cultural Denial. *Population Matters*. Retrieved from https://populationmatters.org/news/2021/11/bill-rees-note-climate-change-and-cultural-denial, 7 February 2022.

Roser, M. (2013). Economic Growth. Retrieved from https://ourworldindata.org/economic-growth, 1 February 2022.

Sen, A. (1999). *Development as Freedom*. New York: Anchor Books.

Socialism Today (2021). Editorial: What Now After the Glasgow Cop-out? *Socialism Today* Issue 254 Dec—Jan 2021/2022. Retrieved from Socialism Today Issue 254 Dec-Jan 2021/2022, 10 February 2022.

The World Bank Group (2019). Rural Population (% of Total Population). Retrieved from https://data.worldbank.org/indicator/SP.RUR.TOTL.ZS, 29 March 2021.

United Nations (n.d.). United Nations Conference on the Human Environment, 5–16 June 1972, Stockholm. Retrieved from United Nations Conference on the Human Environment, Stockholm 1972 | United Nations, 7 February 2022.

United Nations Department of Economic and Social Affairs (UNDP) (2018). News. Retrieved from www.un.org/development/desa/en/news/population/2018-revision-of-world-urbanization-prospects.html, 29 March 2021.

US Department of Commerce (2022). Atmospheric CO_2 at Mauna Loa Observatory. February. Retrieved from https://gml.noaa.gov/webdata/ccgg/trends/co2_data_mlo.pdf, 18 February 2022.

Vale, B. and Vale, R. (2009). *Time to Eat the Dog? The Real Guide to Sustainable Living*. London: Thames and Hudson.

Ward, B. and Dubos, R. (1972). *Only One Earth: The Care and Maintenance of a Small Planet*. New York: Norton.

Weber, M. (1905). (Translation by T Parsons published 1950). *The Protestant Ethic and the Spirit of Capitalism*. New York: Charles Scribner's Sons.

World Bank (2021a). *The Changing Wealth of Nations 2021: Managing Assets for the Future*. Washington, DC: World Bank.

World Bank (2021b). Global Wealth Has Grown, But at the Expense of Future Prosperity. Retrieved from www.worldbank.org/en/news/press-release/2021/10/27/global-wealth-has-grown-but-at-the-expense-of-future-prosperity-world-bank, 7 February 2022.

World Bank Group (2022). GDP Per Capita, (current US$), Retrieved from https://data.worldbank.org/indicator/NY.GDP.PCAP.CD, 8 February 2022.

Worldometer (2022a). World Population by Year. Retrieved from www.worldometers.info/world-population/world-population-by-year/, 18 February 2022.

Worldometer (2022b). World Population. Retrieved from www.worldometers.info/world-population/#table-forecast, 7 February 2022.

11

POPULATION—THE ELEPHANT IN THE ROOM

11.1 Introduction

The human species has been undeniably successful. Its sheer number in terms of population proves this. When the human population reached 7 billion in October 2011, although in both numbers and biomass, there were many more ants and fish than people in the world at that moment, humans were the most numerous large mammals, with cattle estimated at 1.4 billion, sheep 1.1 billion and elephants 0.5 million (Chappell, 2011). Since most of the cattle and sheep are there because the human population wants to eat them, the impact of humanity on the planet is very significant. As an example, poaching has seen a fall of 86% in the numbers of African forest elephants in 31 years "while the population of African savanna elephants decreased by at least 60% over the last 50 years" (IUCN, 2021). Given elephants are intelligent animals, why are they much less successful than human beings?

Robert and Stewart (2018) note that from the Late Pleistocene period, *Homo sapiens* "apparently began to inhabit a huge diversity of other settings across the majority of the Earth's continents". The same authors see *Homo sapiens* as a "generalist specialist" in that the species have occupied a very wide range of habitats, but that populations within these habitats, be they high altitude or desert, have learned to specialise in how they can live successfully. This draws a curious contrast with life in modern cities, which has become more and more homogenous and less specialised in terms of the environments in which the cities are located—at least for the relatively well off. The same types of building are found in almost every large city. The tall, all-glass building is ubiquitous but other types of building have also had an effect. The Bilbao effect, referring to the way Frank Gehry's Bilbao Museum made the Basque city both famous and more prosperous, has led to "countless odd-shaped buildings the world over" in an effort to spur a city to

DOI: 10.4324/9781003045847-13

some form of regeneration (Moore, 2017). Has *Homo sapiens* lost the ability to be a generalist specialist or is it just that the seeming abundance of fossil fuels has made humanity lazy?

The issue of population in a discussion of cities is that cities are viewed as somewhere to house this ever-increasing number of people, as you can have more people apparently living on a smaller area of land in a city, but as has been mentioned earlier, it depends on what land is being measured in an effort to establish how efficiently it is being used. This chapter first looks at the effect of putting more people into cities, by examining homelessness and slums in the 19 cities. It then looks at population growth in different parts of the world and the effect this is having on cities.

11.2 The growth of cities

This section is not intended to be a history of the city. Other authors have given many pages to this topic (Mumford, 1961; Clark, 2013), but it is necessary to investigate the relationship between population growth and city growth, as the world continues to urbanise. Is it the simple increase in human population that drives this global urbanisation or are other factors at work? Buhaug and Urdal (2013) state that urbanisation is linked to three issues: "natural growth, rural–urban migration, and reclassification of areas from rural to urban". Accepting "natural growth" of a city's population, the rural to urban migration is not a simple matter of resources in the rural area becoming insufficient to support its population, although this is a situation that could be exacerbated by climate change (Nawrotzki et al., 2015), but of why people want to migrate and where they want to go. As long ago as 1967, Rose (1967:53) noted that cities that outweigh all others in terms of "population, concentrated commercial and industrial wealth and political strength" tend to attract people and that the resultant growth of these primary cities "begets growth." Rose goes on to note that many such cities are capital cities. So urban growth may be a real problem in such cities, leading to the creation of slum settlements and homelessness. Ooi and Phua (2007) state that both local and national governments "have been narrowly focused on economic growth with the consequent proliferation of slum formation as a housing solution", and this in the face of the knowledge that economic growth will cause city growth. The following section looks at homelessness and slums with a focus on the 19 cities.

11.2.1 Homelessness and slums

National data reveals an increasing trend in homelessness in 24 out of the 27 EU Member States plus the United Kingdom (the exceptions being Finland, Portugal, Croatia and Poland) (Zólyomi et al., 2021:8). It also appears that homelessness is more associated with large cities than small settlements, as is apparent from a 2018 study of homelessness in Germany (Busch-Geertsema et al., 2020).

> In the nine surveyed large cities with a population of over 500,000—among them the city states Berlin, Hamburg and Bremen—approximately 8.6 out of every 1,000 citizens had no home. In cities with a population up to 500,000 the share was 5.3. In the participating small towns and rural municipalities with up to 20,000 citizens, the density was "only" 2.4 homeless people for every 1,000 residents; many small communities reported no cases of homelessness.

The 2018 survey found 2,000 people were homeless in Hamburg, which was nearly twice the figure for 2008 (Zimmermann, 2021). However, comparing data on homelessness between nations and cities is problematic as definitions differ (Chamie, 2020). As an example, as well as those on the street and living in temporary accommodation, the Auckland Council rates as homeless those living in sub-standard permanent accommodation, such as rented garages or overcrowded conditions (Auckland Council, 2021). The official definition of being classified as homeless through living in sub-standard housing is someone living in accommodation lacking in one of six basic amenities: "tap water that is safe to drink; electricity; cooking facilities; a kitchen sink; a bath or shower; a toilet" (Ministry of Housing and Urban Development, 2021). This is similar to the UN definition of a slum (UN-Habitat, 2006), which states slum conditions occur where people live in an urban area in some form of housing that cannot protect them against extreme weather events, where there is not enough living space so that no more than three people share the same room, where there is no easy access to an affordable, safe water supply or to adequate sanitation and where there is no security of tenure. Security of tenure is also an issue in defining homelessness in China (He et al., 2020).

> According to the Ministry of Civil Affairs, China helped 17.7 million vagrants and beggars from November 2012 to June 2019. That estimate constitutes only 1.2 percent of the national population, and it may include the same people counted multiple times. But if we expand the definition of homelessness to incorporate nearly 300 million peasant-workers living a marginal existence on the periphery of cities, the homeless population in China balloons to over 20 percent of the population.

The migrants that come to the cities in China without having a work permit cannot obtain proper accommodation and so "are forced to live in overcrowded and unsanitary conditions" (He et al., 2020). Such accommodation is also prone to be demolished at short notice to make way for middle-class housing, so the residents move on to whatever other accommodation they can find. This would also meet the UN definition of living in slum conditions because of the insecurity of tenure. In contrast, many statistics on homelessness only involve those living on the street or in temporary accommodation, both situations that would also meet the definition of living in slum conditions, since temporary accommodation, such as a shelter for the homeless, is just temporary, even if sanitary.

The highest levels of homelessness are normally associated with the least developed countries and those where wars and natural disasters have displaced people, although local definitions can skew this, as in the case of New Zealand, which has the highest rate of homelessness in the Organisation for Economic Co-operation and Development (OECD) countries (OECD, n.d.). Table 11.1 compares the homelessness rates, where these can be established, for those of the 19 cities which are in the OECD, with national homelessness figures. This is important as Chamie (2020) again makes the point that national rates of homelessness are lower than those of large cities: "while the rate of homelessness [in the USA] is 0.17 percent, the rate in its capital, Washington, DC, is more than seven times higher at 1.24 percent". The fourth column in Table 11.1 indicates where the numbers of homeless include more than the "basic definition" of people living rough, or in emergency accommodation or in accommodation specifically provided for the homeless.

TABLE 11.1 Homeless rates for selected cities in the OECD

City	*Date*	*National (% homeless)*	*Includes more than basic definition*	*Date*	*City (% homeless)*	*Reference*
Amsterdam	2018	0.23	Yes	N/A	N/A	
Auckland	2018	0.86	Yes	2018	1.2	Auckland Council, (2022) Auckland Council (2021)
Copenhagen	2019	0.11	Yes	2021?	0.4–0.6	Christensen& Kirkwood (2022) World Population Review (2022)
Hamburg	2018	0.41	Yes	2018	0.36	Hajek et al. (2021) World Population Review (2022)
New York	2020	0.18	No	2020	0.45*	USICH (2020) United Sates Census Bureau (2021)
São Paulo	2015	0.05	Not provided	2019	0.1	France24 (2021) World Population Review (2022)
Sydney	2016	0.48	Yes	2016	0.5** 3.4***	Australian Bureau of Statistics (2018) Dobson and Montoya (2018)

* Value for New York State: 44% of people live in New York City.

** Value for New South Wales where greater Sydney accounted for 64% of population in 2016 (Australian Bureau of Statistics, 2017).

*** Calculated value for Sydney electorate which includes CBD (Australian Bureau of Statistics, 2021).

While no reliable data could be found for Amsterdam, Table 11.1 shows that, except for Hamburg, the cities, or the states where they are located, have a higher level of homelessness than the national average. The fact the value for Hamburg is lower than the national average may be due to what and what has not been included in the statistics, but it should also be noted that Hamburg has had success with its policies of providing housing for refugees (Spandrel, 2018). The values for both Brazil and São Paulo are low compared to the other OECD figures, but the 2019 Brazilian census figures for the homeless are claimed to be low by those working with the homeless in the city (France24, 2021). Obviously, homelessness is more complicated than arriving in the city believing the streets are paved with gold and finding they are not, since homelessness is also associated with drug abuse and mental illness, but it does seem that increased urbanisation also increases homelessness. This raises the question as to whether increased urbanisation leads to increased slum formation.

The UN-Habitat (2003:v) publication on the world's slums begins with stating that around the start of this century, 32% of the global population lived in slums. More recently, "The most recent global estimates suggest just under 1-in-3 people in urban areas lived in slum households" (Ritchie and Roser, 2018), which implies little change in just over a decade. To look at change over time in more detail, Table 11.2 sets out the national 2005 and 2018 percentages of urban dwellers living in slums. The figures cover those of the 19 cities in countries that are recorded as having urban slum housing by Ritchie and Roser (2018). Table 11.2 also includes information from other sources, where available, on slums in the relevant cities.

With the exception of Buenos Aires, there is a greater percentage of slum dwellers in the large cities than in the nation as a whole. The UN-Habitat report (2003:xxvi) anticipated such an increase in slum formation, at least in some cities:

> More than half of the cities on which case studies were prepared for this Global Report indicated that slum formation will continue (Abidjan, Ahmedabad, Beirut, Bogotá, Cairo, Havana, Jakarta, Karachi, Kolkata, Los Angeles, Mexico

TABLE 11.2 Slum conditions in selected cities

City	*National % urban slums 2005*	*National % urban slums 2018*	*% City population living in slums*	*Date*	*Reference*
Ahmedabad	34.8	35.2	N/A		
Beijing	32.9	24.6	N/A		
Buenos Aires	26.2	14.7	7–10	2020	Machia et al. (2021)
Cairo	17.1	5.2	40	2020	Farid (2020)
Havana	N/A	6.6	N/A		
Jakarta	26.3	30.6	approx. 50	2019	*Jakarta Post* (2019)
Nairobi	54.8	46.5	60–70	2014	Wamukoya et al. (2020)
São Paulo	29.0	16.3	30+	2018	Hoque (2018)

> City, Nairobi, Newark, Rabat-Salé, Rio de Janeiro and São Paulo). A few (Bangkok, Chengdu, Colombo and Naples) reported decreasing slum formation, while the rest reported no or insufficient data on this topic (Durban, Ibadan, Lusaka, Manila, Moscow, Phnom Penh, Quito and Sydney).

However, as noted above there is no agreed definition of what constitutes a slum or those living in slum conditions. The high rate of reported homelessness in New Zealand due to the inclusive nature of what is counted as unacceptable housing conditions has already been mentioned, as have the migrant workers of China. As another example of what is not counted, Dubai does not appear to have a slum problem and has also claimed to have the lowest unemployment rate in the world (Mirandette, 2017). However, it has cohorts of migrant workers that live in overcrowded and unsatisfactory conditions as they attempt to earn enough to send back home (Mirandette, 2017).

The point of looking at homelessness and slum housing is to reveal first that these problems tend to be exacerbated by living in cities, and, second, that as cities grow, these problems do not go away unless there is very serious intervention to try and prevent such inequalities. Singapore is an example of setting the goal of ridding the city of slums (Ministry of National Development, 2019). A 2019 study of homelessness in the city found "between 921 and 1,050 homeless people in Singapore, most of whom were Chinese men" (La, 2020), which is a very small number given the population. Of these, most were in some form of employment but did not earn enough "to afford a place to stay" (La, 2020). Moscow is another city that in the past in the Khrushchev era made great efforts to ensure all citizens were housed (Hatherley, 2015), and although the dwellings may seem very small and unsatisfactory to Western eyes, such policies did ensure everyone could be housed. In a free-market economy, housing is not seen as a right but as a desirable commodity that you can afford if you earn enough, and it seems that increasing numbers of people are finding the latter a problem, as the numbers of homeless, and those living in unsatisfactory and slum accommodation, suggest. The wealth of a city is not an indication of satisfactory housing. Three of the 19 cities fall within the top ten wealthy cities based on private wealth (Beijing 5th, Sydney 9th and Singapore 10th) (World Population Review, n.d.a), and of these, from the discussion above, only Singapore could claim not to have a housing problem.

The issue is that as world population increases so does the rate of urbanisation. We solve one problem, that of more people, by making them live closer together in cities. However, in solving one problem we seem to create others. Nor is housing the only issue that needs to be solved as human populations grow. People need food and energy as well as shelter, and putting people in cities means they cannot supply either of these things for themselves. The increasing complexity of the problems that are raised by increased urbanisation may well lead to some form of societal collapse (Garcia et al., 2021:62). For the homeless, this collapse has already happened.

11.3 Where are national populations growing?

Table 11.3 sets out the rate of population growth of the countries where the 19 cities are located, and the growth of the city where that is known.

It was not possible to find a single source for city growth rates, so the numbers should be taken as indicative rather than absolute. However, apart from Buenos Aires, Hamburg and New York, city growth rates are higher, sometimes much higher, than national population growth rates. Singapore being a city-state obviously has the same growth rate for country and city. Hamburg is losing population at a marginally higher rate than for Germany as a whole. New York is also losing population while

TABLE 11.3 Population growth in the 19 cities

City	*National population 2020*[1]	*National annual population growth rate (%)*[1]	*City annual population growth rate (%)*	*References*
Ahmedabad	1,380,004,385	0.95	6.0 (2021)	Indiaonline (n.d.)
Amsterdam	17,134,872	0.22	0.78 (2021)	Macrotrends (n.d.)
Auckland	4,822,233	0.77	1.9 (2017)	Auckland Growth Monitor (2017)
Beijing	1,439,323,776	0.29	2.12 (2021)	Macrotrends (n.d.)
Buenos Aires	45,195,774	0.89	0.64 (2020)	World Population Review (2022)
Cairo	102,334,404	1.82	2.02 (2021)	Macrotrends (n.d.)
Copenhagen	5,792,202	0.37	0.97 (2021)	Macrotrends (n.d.)
Dubai	9,890,402	0.91	10.7 (2020)	World Population Review (2022)
Hamburg	83,783,942	*−0.02*	*−0.06* (2021)	Macrotrends (n.d.)
Havana	11,326,616	*−0.10*	0.11 (2021)	World Population Review (2022)
Jakarta	273,523,615	1.00	1.46 (2022)	World Population Review (2022)
Kuala Lumpur	32,365,999	1.24	2.54 (2022)	World Population Review (2022)
Moscow	145,934,462	*−0.07*	0.44 (2021)	Macrotrends (n.d.)
Nairobi	53,771,296	2.24	4.0 (2022)	World Population Review (2022)
New York	331,002,651	0.57	*−0.65* (2022)	World Population Review (2022)
Pyongyang	25,778,816	0.40	0.79 (2022)	World Population Review (2022)
São Paulo	212,559,417	0.64	0.88 (2021)	Macrotrends (n.d.)
Singapore	5,850,342	0.79	0.79 (2021)	Macrotrends (n.d.)
Sydney	25,499,884	1.09	3.9 (2021)	Population Australia (2022)

1 World Population Review (n.d.b).

Buenos Aires is growing, but at a slower rate than Argentina. Koop (2021) makes the point that, based on UN population predictions for 2020–2025, the fastest growing cities of the world (none of the top twenty are among the 19 cities considered here) are in Africa (17 cities) and Asia (3 cities). She goes on to say, "By 2025, the world's population will reach over 8.1 billion people. Most of that population growth will be concentrated in cities across Africa and Asia." Table 11.3 reveals the much slower growth rate of the European cities apart from declining Hamburg. Auckland and Sydney also stand out as exceptions in that the growth rates of both are much higher than their European counterparts but that is in part explained by the low population density of both countries, as shown in Table 11.4, where values for growth rates are taken from Table 11.3 and the 2020 values for the population density of the cities are from Table 1.3. The figures for city density for 2021 from the same source have been included for comparison (Demographia, 2021).

The figures for Pyongyang raise a question. The 2015 density was even higher at 16,200 persons/km^2 (Demographia, 2015), the changing values may relate to how the urban area is calculated or the difficulty of obtaining statistics from North Korea. Other discrepancies, such as the slight rise in the population density of Hamburg despite a negative growth rate, come from how the data has been measured,

TABLE 11.4 National and city population densities

City	*National population growth rate 2020 (%)*	*National density 2020 (persons/km²)*[1]	*City population growth rate (%)*	*City density 2020 (persons/km²)*	*City density 2021 (persons/km²)*
Ahmedabad	0.95	428	6.0 (2021)	20,583	21,436
Amsterdam	0.22	411	0.78 (2021)	3,608	3,662
Auckland	0.77	18	1.9 (2017)	2,548	2,701
Beijing	0.29	149	2.12 (2021)	4,658	4,658
Buenos Aires	0.89	17	0.64 (2020)	5,015	5,033
Cairo	1.82	106	2.02 (2021)	9,639	9,845
Copenhagen	0.37	135	0.97 (2021)	2,187	2,840
Dubai	0.91	121	10.7 (2020)	4,004	4,375
Hamburg	*−0.02*	235	*−0.06* (2021)	2,485	2,576
Havana	*−0.10*	103	0.11 (2021)	5,054	5,208
Jakarta	1.00	147	1.46 (2022)	9,756	9,988
Kuala Lumpur	1.24	100	2.54 (2022)	3,831	3,995
Moscow	*−0.07*	9	0.44 (2021)	2,908	3,009
Nairobi	2.24	97	4.0 (2022)	7,065	7,010
New York	0.57	36	*−0.65* (2022)	1,700	1,728*
Pyongyang	0.40	216	0.79 (2022)	12,651	9,613
São Paulo	0.64	25	0.88 (2021)	2,123	6,948
Singapore	0.79	8371	0.79 (2021)	10,981	11,279
Sydney	1.09	3	3.9 (2021)	2,123	2,133

1 World Population Review (n.d.b).

* New York-New Jersey-Connecticut.

which again underlines the state of unknowingness of what is really happening in the world's cities. It may be more helpful to look at what the rise in figures means. Approximately 1,000 people have been added to every square kilometre of Ahmedabad between 2020 and 2021. This means the space per person has dropped from 48.6 m^2 to 46.7 m^2, and this space has to include shares of roads, commercial and industrial buildings, and green space as well as residential space. Against this, Singapore must stand as an example of how urban populations can be housed at very high densities and with very low rates of homelessness, but Singapore is also a rich city-state, although it has problems. Bloomberg Wealth (n.d.) note that "cracks in Singapore's economic model—widening wealth inequality and challenges in buying inexpensive housing—have appeared in many large cities around the world". They also note that these same problems have been exacerbated by COVID-19. Singapore is also a city with an excellent rapid transport system where each journey is subsidised "by more than $1", where eight out of ten households live ten minutes' walk from a station, and where there are concession fares as well as travel vouchers for low-income households (Government of Singapore, 2021). Such a reduction in inequalities does not come from a free-market situation but has to be planned for by governments through a redistribution of money through taxes. Although the link between inequalities and urban living is not the focus of this chapter, the triangle between population growth, inequality and urbanisation seems to be a reality. In her book, Szabo (2016:13) is clear on this point.

> During Malthus' lifetime, overall population in England and Wales grew considerably: from around 6.5 million in 1760 to almost 14 million in 1830, according to the records from the 1831 census. At the same time, poverty dominated the lower strata of society and the increase in food prices made it difficult for the poor to afford food. Growing towns had become home for many new residents, including those with little income, and thus swarmed with homeless families, the sick and individuals with disabilities.

Thus, with an increase in both world population and urbanisation to house this increase, should we also expect an increasing divide between rich and poor? The EF model shows that cities have to bring in goods and services from elsewhere and that this raises the EF of the city. It also means that the cost of these goods and services will reflect the fact they have to be imported. It may well be that increased urbanisation will exacerbate food insecurity and it will not be the wealthy who will suffer. Szabo (2016:104) notes that in 2013 the Food and Agricultural Organisation (FAO) reported that one in eight people was suffering from long-term hunger while in 2008, 35% of adults were overweight. By 2016 the latter was 39% of global adults aged 18+ with 13% classified as obese (WHO, 2022). Being overweight has also been linked to urbanisation and both people being malnourished and people being overweight and obese are occurring in the same country. WHO (2022) note: "While these countries continue to deal with the problems of infectious diseases and undernutrition, they are also experiencing a rapid upsurge in noncommunicable disease risk factors such as obesity and overweight, particularly in urban settings." This again reflects urban inequality and the difference between those who have too little food and those who have too much. Some other impacts of an increasing urban population are briefly explored below.

11.4 Some impacts of burgeoning urban populations

If increased inequality is a possible impact of more people living in closer contact in cites, what else might be affected? Living in close proximity has given rise to the urban heat island (UHI) effect, where higher temperatures are seen in an urban environment compared to the surrounding rural environment. As long ago as 1973, Oke demonstrated a relationship between population and UHI, the latter due to building materials in the urban environment absorbing and storing heat, as well as the lack of evapotranspiration from the limited green areas within the city fabric. UHIs can result in greater water and energy usage, the latter particularly for air conditioning in hot climates (Matthew et al., 2016).

In a study of US city population and crime, Yang et at (2019) found that coming together in cities to produce innovations shared a mechanism with certain crimes, such as robbery, since both depend on social interaction. Whereas the frequency of some crimes, like rape, has a linear relationship with population, others like robbery are enhanced as city populations increase. The fact that the greater social interaction that comes as cities grow has a relationship with some crime is underscored by noting that "Lockdowns in various countries were consistent with a drop in reported street crime" (The Economist Intelligence Unit, 2021:5), although the same source noted that cyber-crime levels increased.

Another issue that comes with an increasing urban population is air pollution and its effect on health. "The driving forces of air pollution include economic development, urbanization, energy consumption, transportation and motorization, as well as increase of urban population" (Chen and Kan, 2008). A 2015 study (Karagulian et al., 2015) found that globally the constituents of urban air pollution from particulate matter with a diameter of 2.5 microns ($PM_{2.5}$) broke down by source into traffic 25%, industrial activities 15%, residential fuel combustion 20%, unspecified from human activities 22% and 18% from dust and salt. They concluded that a reduction in air pollution and its consequences for health required that "solutions to sustainably reduce ambient PM from traffic, industrial activities and biomass burning should urgently be sought".

An increase in traffic through an increase in the urban population can also result in greater congestion. PwC (2019) identified six important areas that underlie urban congestion in US cities. The first is the very reason that people want to be in cities—enhanced economic performance, which unfortunately correlates with vehicle miles travelled (VMT). So without planning for urban mobility through some form of mass transport, congestion on the roads will increase as cities grow. Other factors included the growth in ride hailing, which puts more vehicles on the streets, drivers circling to find parking, and internet buying with the "the resulting increase in single-package deliveries (and smaller vehicle loads), failed deliveries which drive repeat visits, and the high return rate of e-commerce orders" (PwC, 2019).

Health is also a potential problem as urban populations increase. Obesity and becoming overweight can be exacerbated by a more sedentary lifestyle and lack of pleasant places to exercise. The urban poor can suffer from malnutrition and unsanitary living conditions. Increased air pollution can produce health issues and living in closer proximity can increase person-to-person disease transmission.

> Animals and insects serve as efficient vectors for diseases within urban settings and do not discriminate between the rich and poor. The prevalence…of communicable diseases in urban settings, such as tuberculosis (TB), malaria, cholera, dengue, and others, is well established and of global concern.
>
> *Kuddus et al., 2020*

11.5 The future of *Homo sapiens*

If for the majority of the world's growing population the future is urban rather than rural and the same growing population is also living within the limits of the earth's resources, then the city of the future is not going to be the same as the city of today. This is explored further in the conclusion of this book. The issue here is how much can the human population grow and still achieve a reasonable standard of living for all together with what a reasonable standard of living means. Whatever the answer to the last point, which will inevitably differ widely from someone living in a slum in Nairobi to a wealthy plutocrat in Moscow, the rich will have to have less and the poor more. The problem with urbanisation is that it tends to increase the gap between rich and poor, which again points to the need for a new city model if population growth is not to lead to more human tragedy.

References

Auckland Council (2021). Homelessness in Auckland. Retrieved from www.aucklandcouncil.govt.nz/plans-projects-policies-reports-bylaws/our-plans-strategies/auckland-plan/homes-places/Pages/homelessness-auckland.aspx, 7 March 2021.

Auckland Council (2022). Auckland Showing Strong Population Growth. Retrieved from Auckland showing strong population growth - OurAuckland (aucklandcouncil.govt.nz), 7 March 2022.

Auckland Growth Monitor (2017). Auckland Growth Monitor. Retrieved from www.aucklandnz.com/sites/build_auckland/files/media-library/documents/AucklandGrowthMonitor.pdf, 11 March 2022.

Australian Bureau of Statistics (2017). 2016 Census: New South Wales. Retrieved from www.abs.gov.au/AUSSTATS/abs@.nsf/mediareleasesbyReleaseDate/BBEDD77E8EA69ACACA2581BF001E65B2?OpenDocument#:~:text=The%202016%20Census%20counted%207%2C480%2C228,increase%20of%209.8%20per%20cent, 8 March 2022.

Australian Bureau of Statistics (2018). Census of Population and Housing: Estimating Homelessness. Retrieved from www.abs.gov.au/statistics/people/housing/census-population-and-housing-estimating-homelessness/latest-release, 8 March 2022.

Australian Bureau of Statistics (2021). 2016 Census QuickStats. Retrieved from https://quickstats.censusdata.abs.gov.au/census_services/getproduct/census/2016/quickstat/SED10080#:~:text=In%20the%202016%20Census%2C%20there,up%200.6%25%20of%20the%20population, 9 March 2022.

Bloomberg Wealth (n.d.). Singapore Grapples with Keeping the Rich Happy—and Taxing Them Too www.bloomberg.com/news/features/2022-02-16/budget-2022-inequality-is-pushing-singapore-to-consider-more-tax-for-richest, 14 March 2022.

Buhaug, H. and Urdal, H. (2013). An Urbanization Bomb? Population Growth and Social Disorder in Cities. *Global Environmental Change 23*(1), 1–10.

Busch-Geertsema, V., Henke, J. and Steffen, A. (2020). Homelessness in Germany. *European Journal of Homelessness 14*(1), 81–91.

Chamie, J. (2020). As Cities Grow, So Do the Numbers of Homeless. Retrieved from https://archive-yaleglobal.yale.edu/content/cities-grow-so-do-numbers-homeless, 7 March 2022.

Chappell, B. (2011). Along With Humans, Who Else Is in the 7 Billion Club? Retrieved from www.npr.org/sections/thetwo-way/2011/11/03/141946751/along-with-humans-who-else-is-in-the-7-billion-club. 29 October 2021.

Chen, B., and Kan, H. (2008). Air Pollution and Population Health: a Global Challenge. *Environmental Health and Preventive Medicine, 13*(2), 94–101.

Christensen, R.G. and Kirkwood, N. (2022). Homelessness in a Welfare State: Perspectives from Copenhagen. Retrieved from www.humanityinaction.org/knowledge_detail/homelessness-in-a-welfare-state-perspectives-from-copenhagen/#:~:text=Experts%20estimate%20that%20there%20are,the%20Danish%20social%20security%20net, 7 March 2022.

Clark, P. (2013). *The Oxford Handbook of Cities in World History*. Oxford: Oxford University Press.

Demographia (2021). *Demographia World Urban Areas*, 17th Annual Edition. Retrieved from www.demographia.com/db-worldua.pdf, 14 March 2022.

Demographia (2015). *Demographia World Urban Areas*, 11th Annual Edition. Retrieved from www.urbangateway.org/document/demographia-world-urban-areas-2015, 14 March 2022.

Dobson, M. and Montoya, D. (2018). *Homelessness in NSW: Electorate Statistics*. Retrieved from https://apo.org.au/sites/default/files/resource-files/2018-12/apo-nid211701.pdf, 9 March 2022.

The Economist Intelligence Unit (2021). Safe Cities Index 2021. Retrieved from https://safecities.economist.com/, 16 March 2021.

Farid, S. (2020). Can Egypt's Slums Be Replaced by "Safe Zones"? Retrieved from https://english.alarabiya.net/features/2019/03/31/Can-Egypt-s-slums-be-replaced-by-safe-zones-#:~:text=According%20to%20him%2C%20an%20estimated,areas%20across%20Egypt%20are%20unplanned, 10 March 2022.

France24 (2021). Pandemic Pushes New Homeless onto Sao Paulo Streets. Retrieved from www.france24.com/en/live-news/20210828-pandemic-pushes-new-homeless-onto-sao-paulo-streets, 9 March 2022.

Garcia, E., Vale, B. and Vale, R. (2021). *Collapsing Gracefully: Making a built Environment that is Fit for the Future*. Cham: Springer Nature.

Government of Singapore (2021). Building the Best Transport System that is Affordable for All. Retrieved from www.gov.sg/article/building-the-best-transport-system-that-is-affordable-for-all#:~:text=We%20will%20spend%20more%20than,subsidised%20by%20more%20than%20%241, 15 March 2022.

Hajek, A., Bertram, F., van Rüth, V., Kretzler, B., Püschel, K., Heinrich, F. and König, H-H. (2021). Prevalence and Factors Associated with Fear of COVID-19 Among Homeless Individuals During the COVID-19 Pandemic: Evidence from the Hamburg Survey of Homeless Individuals. *Risk Management and Health Care Policy 14*, 2689–2695.

Hatherley, O. (2015). Moscow's Suburbs May Look Monolithic, But the Stories They Tell Are Not. Retrieved from www.theguardian.com/cities/2015/jun/12/moscows-suburbs-may-look-monolithic-but-the-stories-they-tell-are-not, 10 March 2022.

He, H., Su, Z., Zhao, J., Pang, Y. and Wang, Z. (2020). Homelessness and the Universal Family in China. *The American Journal of Economics and Sociology 79*(2), 453–474.

Hoque, L. (2018). Revamping Favelas: Top 10 Facts about Poverty in São Paulo. Retrieved from https://borgenproject.org/tag/slums-in-sao-paulo/, 10 March 2022.

Indiaonline (n.d.). Population of Ahmedabad (2011–2021). Retrieved from www.indiaonlinepages.com/population/ahmedabad-population.html, 11 March 2022.

IUCN (2021). African Elephant Species Now Endangered and Critically Endangered—IUCN Red List. Retrieved from www.iucn.org/news/species/202103/african-elephant-species-now-endangered-and-critically-endangered-iucn-red-list, 29 October 2021.

Jakarta Post (2019). Slums Remain a Fact of Life in Jakarta, Ministry Finds. Retrieved from www.thejakartapost.com/news/2019/05/28/half-of-jakarta-is-slum-ministry-says.html, 10 March 2022.

Karagulian, F., Belis, C.A., Dora, C.F.C., Prüss-Ustün, A.M., Bonjour, S., Adair-Rohani, H. and Amann, M. (2015). Contributions to Cities' Ambient Particulate Matter (PM): A Systematic Review of Local Source Contributions at Global Level. *Atmospheric Environment 120*, 475–483.

Koop, A. (2021). Ranked: The World's Fastest Growing Cities. Retrieved from www.visualcapitalist.com/ranked-the-worlds-fastest-growing-cities/, 11 March 2022.

Kuddus, M.A., Tynan, E. and McBryde, E. (2020). Urbanization: A Problem for the Rich and the Poor? *Public Health Reviews 41*, 1.

La, M-H. (2020). Homelessness in Singapore. Retrieved from https://borgenproject.org/homelessness-in-singapore/, 10 March 2022.

Machia, A., Ferrante, D., Battistella, G., Mariani, J., González, F. and de Quirós, B. (2021). COVID-19 Among the Inhabitants of the Slums in the City of Buenos Aires: A Population-based Study. *British Medical Journal Open 11*. Retrieved from https://bmjopen.bmj.com/content/bmjopen/11/1/e044592.full.pdf, March 10 2022.

Macrotrends (n.d.). Copenhagen, Denmark Metro Area Population 1950–2022, Retrieved from www.macrotrends.net/cities/20894/copenhagen/population, 11 March 2022.

Matthew, A., Sreekumar, S., Khandewal, S., Kaul, N. and Kumar, R. (2016). Prediction of Surface Temperatures for the Assessment of Urban Heat Island Effect Over Ahmedabad City Using Linear Time Series Model. *Energy and Buildings 128*, 605–616.

Ministry of Housing and Urban Development (2021). 2018 Severe Housing Deprivation Estimate—updated. Retrieved from www.hud.govt.nz/research-and-publications/statistics-and-research/2018-severe-housing-deprivation-estimate/, 8 March 2022.

Ministry of National Development (2019). Our City, Our Home: Our Early Struggles. Retrieved from www.mnd.gov.sg/our-city-our-home/our-early-struggles, 10 March 2022.

Mirandette, J. (2017). The Current State of Poverty in Dubai: A Different Kind of Problem. Retrieved from www.borgenmagazine.com/poverty-in-dubai/, 10 March 2022.

Moore, R. (2017). The Bilbao Effect: How Frank Gehry's Guggenheim Started a Global Craze. Retrieved from www.theguardian.com/artanddesign/2017/oct/01/bilbao-effect-frank-gehry-guggenheim-global-craze, 29 October 2021.

Mumford, L. (1961). *The City in History; Its Origins, Its Transformations and Its Prospects*. London: Secker and Warburg.

Nawrotzki, R.J., Hunter, L.M., Runfola, D.M. and Riosmena, F. (2015). Climate Change as a Migration Driver from Rural and Urban Mexico. *Environmental Research Letters 10*(11), 9pp.

OECD (n.d.). HC3.1 Homeless Population. Retrieved from www.oecd.org/els/family/HC3-1-Homeless-population.pdf, 8 March 2022.

Oke, T.R. (1073). City Size and the Urban Heat Island. *Atmospheric Environment* 7, 769–779.

Ooi, G.L. and Phua, K.H. (2007). Urbanization and Slum Formation. *Journal of Urban Health 84*, 27–34.

Population Australia (2022). Sydney Population 2022. Retrieved from www.population.net.au/sydney-population/, 11 March 2022.

PwC (2019). Tackling the Growing Issue of Congestion in Urban Areas. Retrieved from www.pwc.com/us/en/industrial-products/publications/assets/pwc-mobility-insights-congestion.pdf, 16 March 2022.

Ritchie, H. and Roser, M. (2018). Urbanization. Retrieved from https://ourworldindata.org/urbanization, 10 March 2022.

Roberts, P. and Stewart, B.A. (2018). Defining the "Generalist Specialist" Niche for Pleistocene *Homo sapiens*. *Nature Human Behaviour 2*(8), 542–550.

Rose, A.J. (1967). *Patterns of Cities*. Melbourne: Thomas Nelson (Australia) Ltd.

Spandrel, A. (2018). Housing and Integrating Refugees in Hamburg. *Field Actions Science Reports*, Special Issue *18*, 20–25.

Szabo. S (2016). Urbanisation and Inequalities in a Post-Malthusian Context. Heidelberg: Springer.

UN-Habitat (2003). *The Challenge of the Slums*. London: Earthscan.

UN-Habitat (2006). Slums: Some Definitions. Retrieved from https://mirror.unhabitat.org/documents/media_centre/sowcr2006/SOWCR%205.pdf, 8 March 2022.

United Sates Census Bureau (2021). New York State Population Topped 20 Million in 2020. Retrieved from www.census.gov/library/stories/state-by-state/new-york-population-change-between-census-decade.html, 9 March 2022.

USICH (United States Interagency Council on Homelessness) (2020). New York Homelessness Statistics. Retrieved from www.usich.gov/homelessness-statistics/ny/, 9 March 2022.

Wamukoya, M., Kadengye, D.T., Iddi, S. and Chikozho, C. (2020). The Nairobi Urban Health and Demographic Surveillance of Slum Dwellers, 2002–2019: Value, Processes, and Challenges. *Global Epidemiology 2*, 100024.

WHO (World Health Organization) (2022). Obesity and Overweight. Retrieved from www.who.int/news-room/fact-sheets/detail/obesity-and-overweight, 14 March 2022.

World Population Review (2022). World City Population 2022. Retrieved from https://worldpopulationreview.com/world-cities, 7 March 2022.

World Population Review (n.d.a). Richest City in the World. Retrieved from https://worldpopulationreview.com/world-city-rankings/richest-city-in-the-world, 10 March 2022.

World Population Review (n.d.b). 2022 World Population by Country. Retrieved from https://worldpopulationreview.com/, 11 March 2022.

Yagnik, B. (2011). Census Enumeration Counts 10,000 Homeless in City. Retrieved from https://timesofindia.indiatimes.com/city/ahmedabad/census-enumeration-counts-10000-homeless-in-city/articleshow/7606214.cms, 7 March 2022.

Yang, V.C., Papachristos, A.V. and Abrams, D.M. (2019). Modeling the Origin of Urban-Output Scaling Laws. *Physical Review 100*(3), 032306.

Zimmerman, E. (2021). *In Hamburg, five homeless people die in the first week of 2021.* Retrieved from In Hamburg, five homeless people die in the first week of 2021 - World Socialist Web Site (wsws.org), 9 March 2022.

Zólyomi, E., Fuchs, M., Simmons, C., Geyer, L., Sandu, V., Rodrigues, R. and Birtha, M., (2021). *Mapping Trends and Policies to Tackle Homelessness in Europe: A Holistic Approach to Measuring Homelessness Based on Practices of Ten EU Countries.* Vienna: European Centre for Social Welfare Policy and Research.

12
THE INTERNET

12.1 Background

The Internet Society defines the internet as "a world-wide broadcasting capability, a mechanism for information dissemination, and a medium for collaboration and interaction between individuals and their computers without regard for geographic location" (Leiner et al., 1997:2). It arose out of the Defense Advanced Research Projects Agency (DARPA), set up in the United States following the surprise launch of Sputnik, the world's first artificial satellite, by the USSR in 1957, with the aim that the United States would henceforth "be the initiator and not the victim of strategic technological surprises" (DARPA, n.d.). To put it very simply, the internet was a system for linking together many computers so that they could communicate with one another. It first became available for use by the public in the 1990s following the invention of the World Wide Web, which made such intercommunication reasonably straightforward (BBC, 2022).

If you are relatively young, it is probably impossible to imagine how anything could have been done without a computer or access to the internet, both these things have become so universal in our lives. That said, things we do now were quite possible in the past. Before the internet, writing a book like this one involved spending a lot of time in libraries. As an example of how it worked, *The Autonomous House*, a book about designing a house that would operate solely with renewable energy (Vale and Vale, 1975) was written long before the internet and the World Wide Web existed. Writing the book involved a lot of time in the Cambridge University Library, one of the libraries that receives a copy of every publication in the UK under the "Legal Deposit" system, which has been part of English law since 1662 (British Library, n.d.). A typical day of research began by searching for references in the Catalogue Room. The catalogue was a collection of huge books, labelled A to Z, with the entry for each individual book in the whole library

DOI: 10.4324/9781003045847-14

hand-written on a slip of paper, which was pasted into the appropriate place in the catalogue in alphabetical order. When you found the particular book or article you wanted you had to write down its reference number on a scrap of paper (paper and pencils were provided in the Catalogue Room) and then track it down on the shelves of the "Useful Arts" section of the library (as opposed to Fine Arts) where books on Solar Architecture were to be found, along with Agriculture, Knitting and Pottery. Sometimes a chance encounter while searching along a shelf for the reference you wanted would yield another gem.

The whole experience would be familiar to anyone who has seen the Hogwarts' library in the Harry Potter films, but without the magic. The nearest thing to magic came from the cafeteria of the University Library in the early 1970s, which made the world's finest cheese scones to provide a welcome break from the dusty shelves. Around the edges of the library walls were tables and chairs where you could take the books, sit down, read them and copy out quotes and relevant information, all done by hand on paper. Photocopying was available but expensive and hard to organise. The paper notes were taken home to use when writing the book, with pen and paper, prior to typing out a first copy on a typewriter. If you wanted to cut and paste the text you needed scissors, a pot of paste and a brush. Three versions were typed, the last one being the one sent for publishing.

Nowadays it is all much easier and the internet has indeed fulfilled its promise as "a mechanism for information dissemination, and a medium for collaboration and interaction between individuals" (Leiner et al., 1997:2). The problem comes in the kind of information that is disseminated. In the world of pre-internet publishing, and indeed in the world of publishing even now, a would-be author of a non-fiction book submits a written proposal for a book to the publisher, the publisher's representative sends the proposal out to a number of reviewers for comments and then the publisher decides whether or not to invite the author to write the book. The publisher may ask the author to make changes in their proposal based on the reviewers' comments. The whole process of publishing information in this way involves a degree of oversight and interaction before the book finally sees the light of day.

12.2 Social media

On the internet it is easy to post anything at all with little or no oversight whatever. In an attempt to counter some of the problems thrown up by this lack of oversight, New Zealand's Department of Internal Affairs has set up a website called "Keep it real on-line" (DIA, n.d.). One of the most common ways to publish material on the internet is using a range of applications described in general as social media, defined by "(i) profiles for users, (ii) ability for users to constantly upload content, and (iii) ability for users to discuss content and connect with other users" (Ortiz-Ospina, 2019). "Social media platforms [ie Facebook, Twitter, etc.] allow users to have conversations, share information and create web content" (USF, 2022).

In guidelines for university departments wanting to set up social media accounts, the University of South Florida, which has an official university presence on Facebook, Instagram, LinkedIn, Snapchat, TikTok, Twitter and YouTube, advises the following.

> On a personal level, social media allows you to communicate with friends and family, learn new things, develop your interests, and be entertained. On a professional level, you can use social media to broaden your knowledge in a particular field and build your professional network by connecting with other professionals in your industry. At the company level, social media allows you to have a conversation with your audience, gain customer feedback, and elevate your brand.
>
> *USF, 2022*

This is an optimistic view of social media, an opposing view can be seen in a critique by Tom Hodgkinson in *The Guardian*, only four years after Facebook got started. "Why should my relationships be mediated through the imagination of a bunch of supergeeks in California? What was wrong with the pub?" (Hodgkinson, 2008).

Social media are a recent phenomenon; Facebook only began in 2004 (Phillips, 2007) and its use has grown with extreme rapidity. At the start of 2022, social media platforms were used by 4.62 billion people, nearly 60% of the total global population, with Facebook (now called Meta) by far the largest platform with 2.9 billion users (Kepios, 2022). Given the number of people who choose to use social media, these platforms must provide something that the pub does not offer. Hodgkinson (2008) described Facebook as "an enormously successful American business" and as "another uber-capitalist experiment" asking "can you make money out of friendship?" The answer is clearly yes. In 2020 Facebook (Meta) made just over $32 per user, its ARPU (Average Revenue Per User) (McFarlane, 2021). With around 3 billion users that becomes quite a lot of money. McFarlane (2021) points out that advertising is the very purpose of social media's existence: "If you're not paying for the product, the product is you. The real transaction here isn't you receiving enjoyment…but rather, that media company renting your eyeballs to its advertisers."

This is all part of the capitalist dream. An entrepreneur comes up with an idea for making money, launches it successfully and makes lots of money. With the rise of social media, its various platforms have as their primary goal the maximisation of the time you, the user, spend looking at them. They will do anything to ensure that you keep looking for longer, to ensure that you are exposed to as much advertising as possible. To do this, social media platforms use algorithms which predict the sort of things that a user will find interesting and then sends them more of the same in the hope of keeping their attention so that they will look at the advertisements that the advertisers have paid to put on the platform. The algorithms use things like how long or how often you look at something, what you "like" or share with others to predict what you might like to see that will keep you looking (Kim, 2017).

This is all fine if you like looking at films of cute kittens or you want to be in touch with other Morris Minor enthusiasts, but the same algorithms have also been criticised for promoting extremism. The Alliance to Counter Crime Online (ACCO), a programme run by CINTOC (The Center on Illicit Networks and Transnational Organized Crime), a Washington DC-based not-for-profit charity under the "501(c)3" category (Contracts Counsel, 2022), reported that 64% of people who join extremist groups do it as the result of Facebook's recommendation tools. They also stated that while Facebook claims that its AI systems identify and remove 99% of terrorist content, the figure in practice was only around 38% (ACCO, 2020). This makes perfect sense in a world driven by money. A social media company needs to keep people looking at its site, so that they will look at the advertising sold to advertisers. It matters only to keep users' eyes on the site, what attracts them is not an issue as long as it attracts them. You are apparently just as likely to buy a new cell phone while looking for bomb recipes as you are when seeking pavlova recipes.

Another study found that Instagram, founded in 2010 for sharing photos with friends and family and bought by Facebook two years later, has deliberately been promoting extremist content. For example, Instagram users who show interest in anti-vaccination ideas receive recommendations from Instagram's algorithm for QAnon conspiracy theories, misinformation about COVID-19 and for antisemitic content, sometimes combined with adverts for brands (CCDH, 2021).

It is not only extremism that is actively promoted by social media. A recent report from scientists at Cornell University found a 99.9% consensus on human-caused climate change in peer-reviewed scientific literature (Lynas et al., 2021), but this consensus has failed to get through to social media. A recent study created a new Facebook account in the name of a user called Jane. For the purposes of the research, Jane then "liked" a page from a UK organisation campaigning against climate action. Facebook then recommended 27 other things its algorithm thought Jane would enjoy, 26 of which were dedicated to misinformation about climate change (Global Witness, 2022). The study found that users were often directed to more extreme information such as climate-change denial or "chemtrail" conspiracy theories (the theory that the condensation trails of aircraft are chemical sprays for controlling the weather). The study found that "what Facebook's algorithm appears to correctly assume is that a person who already believes one conspiracy theory (say, that the climate crisis is a hoax) is more likely to believe others", and that "Facebook's algorithm is ultimately ensuring that the people most in need of good information are the ones least likely to get it" (Global Witness, 2022)

If you come across a website or a social media platform that is promoting extremism, governments have places where you can report this. In New Zealand for example, Te Tari Taiwhenua, the Department of Internal Affairs, whose slogan is "Building a safe, prosperous and respected nation", provides an online Digital Violent Extremism Form (DIA, 2021). On this you are encouraged to report "articles, images, speeches or videos that promote terrorism or encourage violence; websites made by terrorist or extremist organisations; videos of terrorist attacks and any other content that promotes violent extremism." One of the compulsory requirements, and your complaint will be ignored unless this is filled in, is to provide "Link to primary URL." If you are not familiar with the technical language of the internet, you may not know that this means what most non-specialists call a "web address" the writing along the top of the screen that usually begins with "https" (Walden University, 2021). Clearly it is not easy to complain about extremism if you are not computer savvy. In a similar way, the World Health Organization, in its advice about COVID-19, provides links so that as an individual, you can report COVID-related misinformation to ten different social media outlets (WHO, 2022). Each site has different methods for such reporting and some, like TikTok are far from clear. This does seem a case of reporting crimes to criminals. As seen in the Instagram study above (CCDH, 2021), the fact that you report misinformation may not mean that anything is done about it. This is in spite of Facebook/Meta stating, well before the Centre for Countering Digital Hate study was carried out "We are fully committed to the safety of our community and will continue to expand on this work" (Meta, 2019).

The problem with reporting misinformation is that you have to do it yourself. You as an individual are expected to take your own time to do the reporting, there is no official oversight. It is another part of the privatisation of life. Over the past three decades, services once provided by the state, including transport, education, health care, social care and criminal justice have been privatised in the UK and other countries, leading to, among other things, increased inequality and tax avoidance (Whatfield, 2019). The same thing has happened to individuals. We no longer have milk delivered to the door every morning in glass bottles, with the empties taken away and replaced with filled bottles. In the UK in the 1950s onwards this was usually done by a silent electric milk float. Often the milkman would deliver basic groceries as well. Now we have to drive ourselves to the supermarket to buy it. Public transport to go to work or to the city is no longer available in many places; if we want to travel, we have to buy a car and drive ourselves, taking away the relaxation of reading a book on the bus or taking a nap on the train while someone else does the driving. This privatisation of life is profitable; otherwise it would not happen, and there is more money in selling fifty cars than one bus. It is tempting to see increased urbanisation as one more step in this direction. Living in a high-density environment, it is much harder for people to do some things for themselves and much easier to buy what is necessary, meaning someone, somewhere is making a profit. Tacoli (2012) notes that "Urban residents are more dependent on cash incomes to meet their essential needs than rural residents."

The social media companies carry on promoting misinformation because they have found it makes money for them, but misinformation is not a new problem. Jonathan Swift wrote in 1710, "if a Lie be believ'd only for an Hour, it has done its Work, and there is no farther occasion for it. Falsehood flies, and the Truth comes limping after it" (Swift, 1710). There is nothing new in the internet, it just makes the spreading of misinformation easier, and of course, more profitable.

The problem of misinformation is greater by far than the internet. The environmental writer George Monbiot (2022) sums it up perfectly:

> Nearly everyone...seems to accept that economic growth can and should continue indefinitely on a finite planet...They appear to believe that something they call "the economy" takes priority over our life support systems... They believe that we should pledge unquestioning allegiance to a system we call capitalism even though they are unable to define it, let alone predict where it might be heading.

References

ACCO (2020). *How Social Media Fuels Extremism*. Washington DC: ACCO/CINTOC. Retrieved from www.counteringcrime.org/how-social-media-fuels-extremism, 2 April 2022.

BBC (2022). What Is the World Wide Web. Retrieved from www.bbc.co.uk/bitesize/topics/zkcqn39/articles/z2nbgk7, 18 March 2022.

British Library (n.d.). Legal Deposit. London, British Library. Retrieved from www.bl.uk/legal-deposit, 11 April 2022.

CCDH (2021). *Malgorithm: How Instagram's Algorithm Publishes Misinformation and Hate to Millions during a Pandemic*. London: Center for Countering Digital Hate.

Contracts Counsel (2022). "501(c)3. Contracts Counsel. Retrieved from www.contractscounsel.com/b/501c3#:~:text=A%20501%20%28c%29%20%283%29%20organization%20is%20a%20non-profit,that%20are%20not%20do%20no%20have%20tax-exempt%20status, 2 April 2022.

DARPA (n.d.). About DARPA. Retrieved from www.darpa.mil/about-us/about-darpa, 18 March 2022.

DIA (n.d.). Keep It Real Online. Retrieved from www.keepitrealonline.govt.nz/, 23 March 2022.

DIA (2021). Digital Violent Extremism Form. Wellington, Te Tari Taiwhenua/ Department of Internal Affairs. Retrieved from www.dia.govt.nz/web/submitforms.nsf/digviol, 5 April 2022.

Global Witness (2022). The Climate Divide: How Facebook's Algorithm Amplifies Climate Disinformation. Retrieved from www.globalwitness.org/en/campaigns/digital-threats/climate-divide-how-facebooks-algorithm-amplifies-climate-disinformation/, 11 April 2022.

Hodgkinson, T. (2008). With Friends Like These... *The Guardian*. 14 January. Retrieved from www.theguardian.com/technology/2008/jan/14/facebook, 24 March 2022.

Kepios (2022). Global Social Media Stats. Datareportal. Retrieved from https://datareportal.com/social-media-users, 24 March 2022.

Kim, S.A. (2017). Social Media Algorithms: Why You See What You See. *The Georgetown Law Technology Review 2*(1), 147.

Leiner, B., Cerf, V., Clark, D., Kahn, R., Kleinrock, L., Lynch, D., Postel, J., Roberts, L and Wolff, S. (1997). *Brief History of the Internet*. Reston VA/ Geneva: Internet Society.

Lynas, M., Houlton, B. and Perry, S. (2021). Greater Than 99% Consensus on Human Caused Climate Change in the Peer-reviewed Scientific Literature. *Environmental Research Letters 16*(11), 7pp.

McFarlane, G. (2021). How Facebook (Meta), Twitter, Social Media Make Money from You. *Investopedia.* Retrieved from www.investopedia.com/stock-analysis/032114/how-facebook-twitter-social-media-make-money-you-twtr-lnkd-fb-goog.aspx#:~:text=Key%20Takeaways%201%20The%20primary%20way%20social%20media,advertisers%20who%20use%20the%20platform%20to%20reach%20customers, 1 April 2022.

Meta (2019). Combatting Vaccine Misinformation. Retrieved from https://about.fb.com/news/2019/03/combatting-vaccine-misinformation/, 11 April 2022.

Monbiot, G. (2022). Organised Lying. *The Guardian.* 30 March. Retrieved from www.monbiot.com/2022/04/03/organised-lying/, 11 April 2022.

Ortiz-Ospina, E. (2019). The Rise of Social Media. *Our World in Data.* Retrieved from https://ourworldindata.org/rise-of-social-media, 24 March 2022.

Phillips, S. (2007). A Brief History of Facebook. *The Guardian* 25 July. Retrieved from www.theguardian.com/technology/2007/jul/25/media.newmedia, 24 March 2022.

Swift, J. (1710). The Examiner, Number 15. Retrieved from www.bartleby.com/209/633.html, 7 April 2022.

Tacoli, C. (2012). Urbanization, Gender and Urban Poverty: Paid Work and Unpaid Carework in the Cities. *Urbanization and Emerging Population Issues. Working Paper* 7. Retrieved from https://pubs.iied.org/10614iied, 13 May 2022.

USF (2022). Introduction to Social Media. University Communications and Marketing. Retrieved from www.usf.edu/ucm/marketing/intro-social-media.aspx, 24 March 2022.

Vale, B. and Vale, R. (1975). *The Autonomous House London.* London: Thames and Hudson.

Walden University (2021). Internet Basics: Web Addresses. Retrieved from https://academicguides.waldenu.edu/c.php?g=81413&p=519805, 7 April 2022.

Whitfield, D. (2019). *Public Alternative to the Privatisation of Life.* Nottingham: Spokesman Books.

WHO (2022). How to Report Misinformation Online. Retrieved from www.who.int/campaigns/connecting-the-world-to-combat-coronavirus/how-to-report-misinformation-online, 21 March 2022.

13
ICONS IN THE CITIES

13.1 Introduction

The original idea for this chapter was to identify new architectural icons in the 19 cities and then focus on the importance of these buildings in shaping the cities' identity, including the environmental impact of new architecture. Although the identification of icons, being a subjective matter, is not a simple task, finding sufficient data to assess detailed environmental impact, including energy and water consumption, proved impossible, despite the best efforts of two interns.[1] Consequently, this chapter focusses on the relationship between the production of new architectural icons as the materialisation of current cultural values, and how these new buildings are shaping and promoting cities to attract businesses and people. Ho (2006) has suggested that the major clientele for these new self-proclaimed icons are corporations. His argument is simply that cities may need these iconic buildings to attract multinational companies and, therefore, capital. The result seems to have become a competition of forms and extravagant shapes.

McDonell and Burton (2006) identified that one of the purposes of tourism is culture. They divided cultural tourism into two groups. The first, conceptual cultural tourism, implies the movement of persons to a cultural attraction away from their usual place of residence, with the aim of gathering new information and experiences to satisfy their cultural needs. The second, technical cultural tourism, considers all the movements of people to specific cultural attractions, such as heritage sites, artistic and cultural manifestations, and arts and drama outside of their usual place of residence. The many types of cultural attractions include cultural and historical monuments, heritage attractions, music, opera, religion and museums. Cultural functions are unavoidably related to architecture since every piece of architecture or built area open to the public as a cultural attraction is designed for

DOI: 10.4324/9781003045847-15

some cultural function (Dai et al., 2018). The architecture is also an initial cultural attraction on its own, attracting tourists and shaping the city's image (Brizotti-Pasquotto and Medrano, 2014).

13.1.1 Architecture and the cultural landscape

Hudman and Jackson (1999) identified architecture as the most visible aspect of the cultural landscape. A city's architecture is also one of the main attractions for its visitors (Shulman, 2012). An informal competition of forms between cities seems to be taking place, aiming to make a competing city more appealing and capable of attracting visitors (Vukadinović, 2011). These new self-proclaimed architectural icons, besides materialising current cultural consumerism (Sklair, 2017) aim to attract companies and their investments through the modification of a city's skyline often with showy buildings that are also there to attract tourists. If cultural tourism is indeed relevant and these icons are manifestations of our cultural moment (McDonell and Burton, 2006; Sklair, 2017), the competition between cities using these new icons to attract business and people is understandable but environmentally questionable.

McNeill (2009) called this new, iconic, privately developed, architecture signature architecture. This is not a new phenomenon. Examples range from Frank Lloyd Wright's Johnson Wax Building (opened 1939) and SOM's Sears Tower (1973) to Foster and Partners' Hong Kong and Shanghai Bank (completed 1986) or Kenzo Tange's Fuji Television building in Tokyo (1997) (Pearman, 2002). It is important to notice the shift post World War II in the production of icons. Once the production of architectural icons was mostly backed by the public sector, such as the parliamentary building known as the Beehive in Wellington (1979), but this has shifted to a more complex capitalisation structure that is often private (Sklair, 2017). Curtis (1996) has identified that, throughout history, monumental architecture has been employed to embody the values of the dominant ideology. Krieger (1999) noticed a contradiction between the power of a dense New York skyline, against the myth of the isolated tower in Europe that is often a church, and what both symbolise. He described how the West German urban planner from the 1950s, Rudolf Hillebrech, considered a skyline like New York interesting, but that it expressed a commercial spirit that was the opposite of the community values expressed in the more modest German skylines of the time. This difference also expresses a difference in funding of the buildings that make the skyline. Shapely (2011) identified that cities in England have adopted entrepreneurial strategies to redevelop city centres using public–private partnerships, and Fainstein (2001) studied the participation of developers in the real estate market in shaping London and New York. From a more theoretical approach (Harvey, 1989; Marcuse and Madden, 2016; Sadri and Zeybekogly, 2021; Brenner et al., 2011) to an environmental approach (Chicca, 2013), studies have indicated how economic agents have been shaping cities. Jencks (2006) finalises this by saying that architects and their commercial products are now simply competing for attention.

13.2 Office buildings as new icons

A 2011 study suggests higher prices are possible for the rental and sales of office buildings that have been designed by famous architects (Fuerst et al., 2011). McNeill (2009) identified the strategy of developers for minimising risks. Developers, particularly property trusts, try to reduce risks by pre-letting buildings to large corporations, which often demand architectural aspects that differentiate them from competitors' buildings. As Larry Malcic, a director of HOK (formerly Hellmuth, Obata + Kassabaum) suggested, developers have identified that architecture matters and they are after something iconic (whatever iconic may mean) to set their buildings apart (McNeill, 2009). Sklair (2017), besides classifying architectural icons, also identified the use of icons as the materialisation of the culture-ideology of consumerism. He goes on to state that the promotion of this new architectural iconicity is not limited to professional architecture but is an effort made by all media. Given the developers' marketing strategy of using famous architects to create these icons, and the media promotion of these new buildings, it is understandable that cities want to use these, often monumental, constructions to symbolise development.

Kaika (2010) has suggested an interesting argument about the transformation of London's skyline. Her argument is based on the use of new built icons, such as 122 Leadenhall Street, the Cheese Grater (by Rogers Stirk Harbour + Partners) and 30 St Mary Axe, the Gherkin (by Foster + Partners). She believes that these icons are an effort to materialise the economic shift from a British banking élite, more loyal to local traditions (in this case the skyline), to international corporations, less loyal to such formal limitations. Sklair and Struna (2013) investigated further the concept of the transnational capitalist class (TCC), proposed by Sklair (2001) and its four components: the corporate, political, professional and consumerist. The corporate aspect is simple to understand; it refers to the player, often an investor or a developer, who is interested in profit, shaping well-performing brands into iconic commodities. The political aspect is not as direct. Among its more complex characteristics, one important factor is to provide a narrative for incorporating these new icons into a nationalist image or feeling. Another purely political aspect refers to the demand of certain authorities for the "wow factor" in new buildings (Jencks, 2006). This leads to the acceptance of commercial-private office building as a central player in the shaping of a city's new image or aspiration. The professional fraction provides the professional support for the execution of these new icons, such as the architects and designers who provide the narrative to validate them, by promoting the new building as a "new icon" for a certain city. Sklair and Gherardi (2012) have explored some of the websites of top architectural offices to verify that indeed the offices call their own designs icons. They went further to identify how often these self-proclaimed icons were shown in the media, and which architects were more often mentioned. The final component is the consumerist, which is the hardest to define. It refers to the capacity to transform these icons into consumable items or to use them as leverage for the consumption of something. In trying to attract companies and visitors, cities ultimately want to sell themselves. However, it is hard to see these private icons as reflecting the shared memories, identity and solidarity of societies, as suggested by Ho (2006). There is, however, some evidence that, at least to some extent, some people may be convinced that these buildings may indeed be or become icons. In a survey carried out by Kaika (2010) in London with city workers, the Gherkin was considered an icon for more people than St Paul's Cathedral, the Bank of England, the Tower of London, London Bridge or Tower Bridge. In fact, it was pointed out as the main icon of London. Although not definitive because of the sample size and only related to London, this suggests that these self-proclaimed icons have the potential to become city icons.

13.3 The Bilbao effect

The production of office buildings as icons has both similarities and differences with what happened in Bilbao. The Bilbao effect is the success story of the urban regeneration of the city of Bilbao in the Basque Country in Spain through the apparently inexplicable and unprecedented success of the Guggenheim Museum Bilbao (GMB) in catalysing the regeneration of Bilbao's urban area (Franklin, 2016; Plaza and Haarich, 2013) (see Chapter 11). Regardless of the importance of the GMB in regenerating the city, the museum itself was successful. Originally planned to receive 400,000 visitors a year, it received a million in its first year and has maintained that visitation level ever since (Areso, 2007), paying back the cost of the building in five years and overall investment in 10 years (Franklin, 2016). It became a great contributor of GDP to the city, from 2006 adding $35 million annually to Basque public funds. When the GMB turned 20 years old, it had contributed $4.3 billion to the GDP of the Basque Country and generated $695 million in additional tax revenue for the region (McGivern, 2017).

It is hard to deny that the GMB was not a success from the investment point of view, but it has failed to restructure the region's uneven development and inequality (Rodriguez et al., 2001) or to solve the city's unemployment, poverty and gentrification problems (Vicario et al., 2003). The fact is the benefits created in Bilbao are not only from the museum. It was a more complex urban plan that involved place-marketing campaigns that altered the city's image through investments in large-scale projects. Here is where there are some similarities with the new self-proclaimed icons. Several important architects, besides GMB designer Frank Gehry, were involved in projects in Bilbao, including Norman Foster, Cesar Pelli and Arata Isozaki (Vicario et al., 2003).

The differences between the GMB and the new iconic buildings are in the use of the building as a museum as opposed to the normal office use and the lack of a developer figure behind the museum. However, although developers were not directly involved with the museum, the Bilbao effect has generated many opportunities for them, and the process of gentrification in many areas of the city is related to the shift in the city's dynamics (Vicario et al., 2003). There are several similarities between GMB and the office icons. First, and perhaps most obvious, is the use of an internationally famous architect as the designer, followed by the use of the building's image to promote the city. However, the most remarkable feature of the GMB may not be in Bilbao, but in the cities that have tried to replicate the apparent success of the museum, either by designing unusually shaped museums, such as the alien-blob of the Kunsthaus Graz, designed by Peter Cook and Colin Fournier, or by simply using the new redesigned skyline to promote the success of the city and attract visitors. While it is hard to measure the environmental impact of these new museums and office buildings, particularly regarding the operational and construction impact, there is a fair amount of data regarding the environmental impact of the tourism they create.

13.4 The impact of tourism

The Ecotourism website has a catchphrase that, they claim, perfectly summarises sustainable tourism: Take only photos and leave only footprints (Ecotourism World, 2020). This does not, however, match with what recent data shows about the environmental impact of tourism. The *International Tourism Highlights* (2020) show that 94% of tourism happens via air (59%) or roads (35%), leaving rail and water at a mere 6% (UNWTO, 2021). As a result, it is probable that the majority of tourists first arrive in urban areas, where most of the airports are located. In fact, between 2012 and 2016, in Europe the number of beds in urban areas increased by 14.2% fuelled by improvements in air connectivity (Boivin and Tanguay, 2019). The UNWTO (2021) suggests that export earnings from international tourism are a relevant source of income for many destinations, tourism being the third-largest export category after fuels and chemicals. The same report indicates that Europe alone receives more than half the world's international arrivals, with only 11% of travel being for business purposes while the remaining 83% breaks down into leisure (51%), and travel to visit friends or for religion or health (28%). The impact of flights, including freight, is around 3.5% of total global CO_2 emissions (Our World in Data, 2020). Lenzen et al. (2018) gave a more detailed estimate of the environmental impact of tourism, in which transport accounted for 49% of total emissions, followed by consumption of goods (12%) and food (8%), with the remaining 30% distributed among services, accommodation, construction and commodities. There is, however, a more concerning figure related to tourism presented in the *Transport-related CO_2 Emissions of the Tourism Sector—Modelling Result Report* by the World Tourism Organization of the United Nations (UNWTO). This report presents data from before the pandemic, meaning the projections may require some adjustment. Regardless of the effect of the pandemic, the numbers are not favourable for a less unsustainable tourism market (Figure 13.1).

The same report presents projections regarding the increasing share of tourist air travel in all modes (Figure 13.2) and in interregional tourism (Table 13.1).

The concern rests in the fact that in all scenarios the largest contributor to greenhouse gas (GHG) emissions in tourism will increase through more flights. At this moment, there is no technology to decarbonise air travel (Our World in Data, 2020). Although some companies promise zero-emission commercial flights by

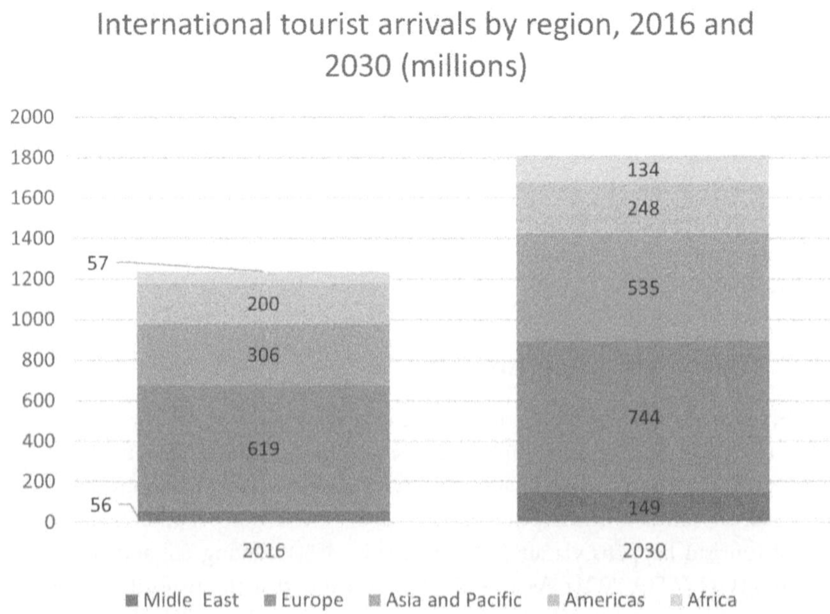

FIGURE 13.1 International tourist arrival by region in 2016 and projection for 2030 (millions) (UNWTO, 2019)

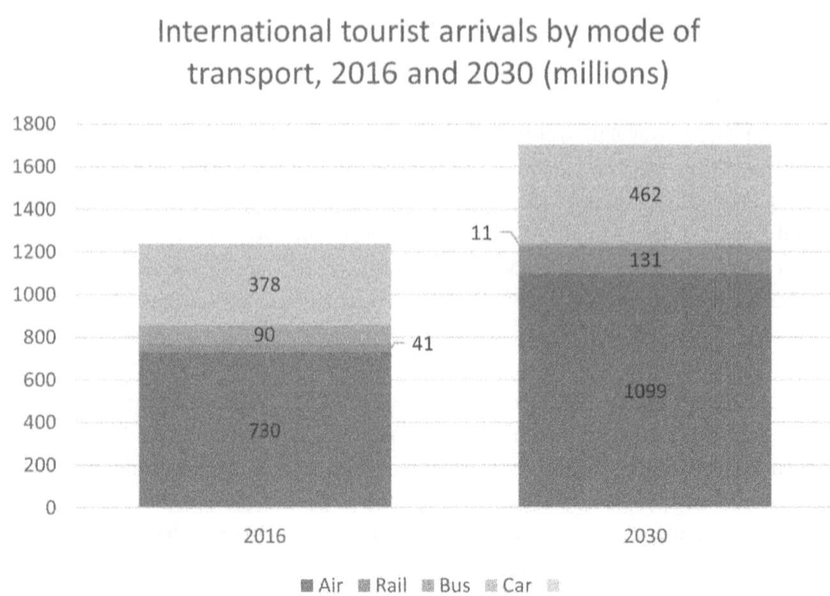

FIGURE 13.2 International tourist arrivals by mode of transport in 2016 and projection for 2030 (millions) (UNWTO, 2019)

TABLE 13.1 Intraregional tourist arrivals by mode of transport in 2016 and projection for 2030 (millions) (UNWTO, 2019)

		Car	*Bus*	*Rail*	*Air*
Africa	2016	3.0	0.2	0	26.3
	2030	2.7	0.2	0	40.4
Americas	2016	63.0	7.0	2.7	76.3
	2030	68.7	7.4	1.4	116.0
Asia and the Pacific	2016	76.2	14.6	1.4	156.2
	2030	82.5	13.8	6.5	245.3
Europe	2016	230.5	67.6	26.3	204.1
	2030	304.3	89.1	94.7	326.2
Middle East	2016	4.2	2.5	0	13.8
	2030	2.3	7.5	0	17.0

2035 using hydrogen (Airbus, 2022), others are less optimistic and sceptical about the capacity for replacing conventional fuel with hydrogen (Davis et al., 2018), as there are considerable technical challenges to solve (Cecere et al., 2014; Fulton et al., 2015).

The projections from UNWTO suggest an increase in the number of passengers and an increase in the share of flights in total tourism transportation. This puts enormous pressure on any idea of achieving sustainable tourism. Given that transportation is roughly half the emissions from tourism, the projected increase from 730 million to 1,099 million passengers in 2030 would require the airline industry to become at least 33% more efficient to keep the emissions at the level we have today, which is already not sustainable. Even considering the best possible scenario of a technology for zero-emissions flights in 2035, it would require many years for all the fleets around the world to be replaced. Although sorting out the transportation emissions would be a big achievement, this would still only address half of tourism's environmental impact. The increase in people travelling will inevitably offset all the efforts from the tourism industry to make the activity less unsustainable. The tragedy is that although there have been travel restrictions during the pandemic, it is unlikely that the same restrictions would be applied for environmental reasons.

13.5 Conclusion

The creation of new urban icons, as representations of the consumerist culture (Sklair, 2017), is both an architectural symbol of our times and a worldwide phenomenon. Because of their size and geographical spread, their environmental impact cannot be ignored. This environmental impact goes beyond their physical footprint, their construction and operation. It is also related to the impact on the image of cities in attracting visitors and business, and the direct impact on the urban fabric. As the market reshapes our cities following the principle of profit by adding new icons, it becomes hard to distinguish how different political systems, whether democratic or not or left- or right-wing, plan to include the environment in the urban equation. It seems clear that ensuring that these new icons are energy efficient is not enough. Rebranding a city is not unusual and requires political will and strong advertising efforts. Amsterdam is an example. It shifted its 1970s image from being associated with cheap drinks, commercial sex and readily available drugs (Kavaratzis and Ashworth, 2007) to promoting historic canal houses, museums, churches and metropolitan architecture (Dai et al., 2018). Although it is possible to see city councils using these new icons to promote themselves, there is little information connecting these new icons to an increase in tourism, besides some rare examples like the GMB. There is less publicly available information for measuring the environmental impact of these icons or even for assessing the need for these buildings. These new icons seem to be part of rebranding and promotion efforts. What can be seen is a competition between cities that aims to attract companies, show off prosperity and gather more visitors. If a city is perceived as being prosperous through having more iconic buildings, it is more likely to attract companies, in turn increasing prosperity, and attracting more visitors, thereby generating more capital for more investment. Although some of these buildings may have solid environmental credentials, the environmental impact is not from the buildings themselves. Apart from visitors, such impact might come from urban changes through the gentrification process. The environmental damage from icons exists but cannot be properly measured. The perception that a building itself is an entity isolated within the urban fabric is convenient for promoting supposed environmental credentials, and an unusual shape may even help with this perception. A new self-proclaimed iconic building, which may stand out because of its differences from the city skyline, might have a better environmental performance, but it will simply be offset by the increase in people coming to see it.

This raises discussion regarding the complexity of the allocation of the environmental impact of tourism. Imagine that an individual takes a flight from New York to Paris, and from Paris, this tourist rents a car and visits different cities in Europe. The CO_2 emissions from this travel may be very hard to distribute.

A city that is trying to reduce its emissions to become carbon neutral should account for the CO_2 emissions from its visitors. The total impact, including tourism, caused by these icons should make those worried about the environment question the need for these kinds of buildings and the promotional efforts of the cities that house them.

Note

1 Hannah Bendull from the United States and Karina Iurkova from Russia.

References

Airbus (2022). ZEROe—Zero emission—Airbus. Retrieved from www.airbus.com/en/innovation/zero-emission/hydrogen/zeroe, 16 May 2022.

Areso, I. (2007). Bilbao's Strategic Evolution. From Industrial to the Post-Industrial City. Osaka Transformation Conference. Retrieved from www.riss.osaka-u.ac.jp/jp/events/point/OSAKATransformation%20of%20Bilbao%20.pdf, 15 May 2022.

Boivin, M. and Tanguay, G.A. (2019). Analysis of the Determinants of Urban Tourism Attractiveness: The Case of Québec City and Bordeaux. *Journal of Destination Marketing & Management 11*, 67–79.

Brenner, N., Marcuse, P. and Mayer, M. (2011). *Cities for People, Not for Profit: Critical Urban Theory and the Right to the City*. Abingdon: Routledge.

Brizotti-Pasquotto, G. and Medrano, L. (2014). Iconic Buildings and City Marketing: The Central Area of São Paulo. *Bitácora Urbano Territorial 24*(2), 21–30.

Cecere, D., Giacomazzi, E. and Ingenito, A. (2014). A Review on Hydrogen Industrial Aerospace Applications. *International Journal of Hydrogen Energy 39*(20), 10731–10747.

Chicca, F. (2013). Sustainable Urban Form. In Vale, R. and Vale, B. (eds). *Living within a Fair Share Ecological Footprint* (pp. 318–330). Abingdon: Routledge.

Curtis, W. (1996). *Modern Architecture Since 1900* (3rd ed.). London: Phaidon.

Davis, S.J., Lewis, N.S., Shaner, M., Aggarwal, S., Arent, D., Azevedo, I.L., Benson, S.M., Bradley, T., Brouwer, J., Chiang, Y.M., Clack, C.T.M., Cohen, A., Doig, S., Edmonds, J., Fennell, P., Field, C.B., Hannegan, B., Hodge, B.M., Hoffert, M.I... Caldeira, K. (2018). Net-zero Emissions Energy Systems. *Science 360*(6396), 9.

Dai, T., Zhuang, T., Yan, J. and Zhang, T. (2018). From Landscape to Mindscape: Spatial Narration of Touristic Amsterdam. *Sustainability 10*(8), 2623–2643.

Ecotourism World (2020). Sustainable Travel: Take Only Photos, Leave Only Footprints. Ecotourism World. Retrieved from https://ecotourism-world.com/sustainable-travel-take-only-photos-leave-only-footprints/, 14 May 2022.

Fainstein, S.S. (2001). *The City Builders: Property, Politics, and Planning in London and New York*. New York: Blackwell.

Franklin, A. (2016). Journeys to the Guggenheim Museum Bilbao: Towards a Revised Bilbao Effect. *Annals of Tourism Research 59*, 79–92.

Fuerst, F., McAllister, P. and Murray, C.B. (2011). Designer Buildings: Estimating the Economic Value of "Signature" Architecture. *Environment and Planning A 43*(1), 166–184.

Fulton, L.M., Lynd, L.R., Körner, A., Greene, N. and Tonachel, L.R. (2015). The Need for Biofuels as Part of a Low Carbon Energy Future. *Biofuels, Bioproducts and Biorefining 9*(5), 476–483.

Harvey, D. (1989). From Managerialism to Entrepreneurialism: The Transformation in Urban Governance in Late Capitalism. *Geografiska Annaler. Series B, Human Geography*, *71*(1), 18.

Ho, K.C. (2006). Where Do Community Iconic Structures Fit in a Globalizing City? *City 10*(1), 91–100.

Hudman, L.E. and Jackson, R.H. (1999). *Geography of Travel and Tourism*. New York: Delmar.

Jencks, C. (2006). The Iconic Building is Here to Stay. *City 10*(1). 3–20.

Kaika, M. (2010). Architecture and Crisis: Re-inventing the Icon, Re-imag(in)Ing London and Re-branding the City. *Transactions of the Institute of British Geographers 35*(4), 453–474.

Kavaratzis, M. and Ashworth, G.J. (2007). Partners in Coffeeshops, Canals and Commerce: Marketing the City of Amsterdam. *Cities 24*, 16–25.

Krieger, P. (1999). The American Impact on Western Europe: Americanization and Westernization in Transatlantic Perspective. Conference at the German Historical

Institute Washington. Retrieved from https://perspectivia.net/servlets/MCRFileNodeServlet/ploneimport3_derivate_00003254/21-33mauch_europe.pdf, 16 May 2022.
Lenzen, M., Sun, Y.Y., Faturay, F., Ting, Y.P., Geschke, A. and Malik, A. (2018). The Carbon Footprint of Global Tourism. *Nature Climate Change 8*(6), 522–528.
Marcuse, P. and Madden, D. (2016). *In Defence of Housing*. London: Verso Books.
McDonell, I. and Burton, C. (2006). The Marketing of Australian Cultural Tourist Attractions: A Case Study from Sydney. *International Cultural Tourism 16*, 16–25.
McNeill, D. (2009). *The Global Architect*. London: Taylor and Francis.
McGivern, H. (2017). Twenty Years on: How the Guggenheim Bilbao Came of Age. *The Art Newspaper*. Retrieved from www.theartnewspaper.com/2017/11/24/twenty-years-on-how-the-guggenheim-bilbao-came-of-age, 12 May 2022.
Our World in Data. (2020, October 22). Climate Change and Flying: what Share of Global CO_2 Emissions Come from Aviation? Retrieved from https://ourworldindata.org/co2-emissions-from-aviation, 13 May 2022.
Pearman, H. (2002). *Contemporary World: Architecture*. London: Phaidon.
Plaza, B. and Haarich, S.N. (2013). The Guggenheim Museum Bilbao: Between Regional Embeddedness and Global Networking. *European Planning Studies 23*(8), 1456–1475.
Rodríguez, A., Martínez, E. and Guenaga, G. (2001). Uneven Redevelopment: New Urban Policies and Socio-Spatial Fragmentation in Metropolitan Bilbao. *European Urban and Regional Studies 8*(2), 161–178.
Sadri, H. and Zeybekoglu, S. (2021). *Cities Without Capitalism*. Abingdon; New York: Routledge.
Shapely, P. (2011). The Entrepreneurial City: The Role of Local Government and City-Centre Redevelopment in Post-War Industrial English Cities. *Twentieth Century British History 22*(4), 498–520
Sklair, L. (2001). *The Transnational Capitalist Class* (Vol. 17). New York: Blackwell.
Sklair, L. (2017). *The Icon Project*. Oxford: Oxford University Press.
Sklair, L. and Struna, J. (2013). The Icon Project: The Transnational Capitalist Class in Action. *Globalizations 10*(5), 747–763.
Shulman, G. (2012). Connecting People with Architecture Architecture's New Role in Developing Countries. *6th Conference of the International Forum on Urbanism (IFoU): TOURBANISM* (pp. 1–11). Barcelona: International Forum on Urbanism.
UNWTO (2019). *Transport-related CO_2 Emissions of the Tourism Sector—Modelling Results*. Madrid: World Tourism Organization (UNWTO).
UNWTO. (2021, January 19). *International Tourism Highlights, 2020 Edition*. Madrid: World Tourism Organization (UNWTO).
Vicario, L., Manuel, P. and Monje, M. (2003). Another "Guggenheim Effect"? The Generation of a Potentially Gentrifiable Neighbourhood in Bilbao. *Urban Studies 40*(12), 2383–2400.
Vukadinović, I. (2011). Architecture in Tourism, Case of Copenhagen, Visitors' Perspective. Master's Thesis. Helsingborg: Lunds Universitet.

14

PANDEMIC

14.1 COVID-19 and this book

This book was written entirely during the COVID-19 pandemic. A pandemic is a widespread outbreak of an infectious disease that can increase morbidity (illness) and mortality (death) over a region and cause significant economic, social and political disruption (Madhav et al., 2017). The definition of a pandemic rather than a disease comes from the World Health Organization (WHO), a specialist agency of the United Nations (UN). The WHO declared COVID-19 to be a pandemic on 11 March 2020 (WHO, 2020a). Although data collection for this book started in late 2019, it was written during the pandemic.

The pandemic not only affected the data collection for the book, as the international interns involved could not come to New Zealand (the authors' homeland), but also some of the social and economic aspects that directly relate to the environment. A direct relationship between the pandemic and the denial of climate change and environmental impact became apparent through the debates about vaccines and social restrictions. The anti-vax movement, anti-lockdown protesters and COVID-19 deniers share some similarities with those still denying the environmental catastrophe that we and future generations will face. A relationship between economic growth, population growth, deforestation and pandemics also emerged.

14.2 The effect of COVID-19

Virtually all the countries and cities assessed in this book are being affected by COVID-19. Because of how the virus spreads, each country is in a different stage of the pandemic. In order to compare the situation of each city at the time of writing this book, albeit roughly, the American Bureau of Consular Affairs and Smartraveller from the Australian federal government were consulted, as both

DOI: 10.4324/9781003045847-16

TABLE 14.1 Traveller information for the 19 cities in response to the January 2022 Omicron surge

Cities	*US advice*	*Australian advice*
Ahmedabad	Level 2	Exercise a high degree of caution
Amsterdam	Level 4	Exercise a high degree of caution
Auckland	Level 1	Exercise a high degree of caution
Beijing	Level 3	Exercise a high degree of caution
Buenos Aires	Level 3	Exercise a high degree of caution
Cairo	Level 3	Reconsider your need to travel
Copenhagen	Level 4	Exercise a high degree of caution
Dubai	Level 1	Exercise a high degree of caution
Hamburg	Level 4	Exercise a high degree of caution
Havana	Level 2	Reconsider your need to travel
Jakarta	Level 1	Exercise a high degree of caution
Kuala Lumpur	Level 3	Exercise a high degree of caution
Moscow	Level 4	Reconsider your need to travel
Nairobi	Level 3	Exercise a high degree of caution
New York	N/A	Exercise a high degree of caution
Pyongyang	Unknown	Do not Travel
São Paulo	Level 3	Reconsider your need to travel
Singapore	Level 4	Exercise a high degree of caution
Sydney	Level 3	N/A

institutions provide traveller information regarding COVID-19 in different parts of the world (Table 14.1). The US advice classified countries by four levels (Level 1—Exercise normal precautions; Level 2—Exercise increased caution; Level 3—Reconsider travel; Level 4—Do not travel). The Australian version also has four categories, with the same recommendations.

The American version is more restricted regarding European countries, probably reflecting the impact of the Omicron variant, while the Australian version is somewhat more lenient despite Europe now (in the first week of January 2022) being a hotspot for COVID-19 (*The New York Times*, 2022). At the time of writing, countries are fighting a fourth or fifth wave and the Omicron variant. While some cities are already easing restrictions, as in Copenhagen where cinemas and music venues have reopened (Reuters, 2022), places like Auckland are preparing for the arrival of the Omicron variant that has not yet been circulating in the community. Early in 2020, during the first months of the pandemic, weapons against the virus were similar all over the world—movement restriction and social distancing—leading to similar responses globally.

14.3 Climate change and pandemics

The expansion of human habitats to areas covered by forests increases the risk of zoonotic diseases (diseases transmitted to humans from animals) with risk of reaching the pandemic level (Brancalion et al., 2020). Before COVID-19, Allen et al. (2017), Rulli et al. (2017) and Rohr et al. (2019) had already pointed out in different research areas the imminent risk of a pandemic arising from deforestation. Andrée et al. (2019) suggested that although development gradually improves the efficiency of consuming natural resources, increased efficiency alone is not sufficient to offset growth in scale, meaning that, despite all efforts, deforestation will increase in the coming years. Their research confirms the findings from Cuaresma et al. (2016) who framed a link between economic development and overexploitation of natural resources including forests. The literature connecting deforestation and pandemics is vast. While there are risks for zoonotic diseases from eating wild animals, the risk of a new pandemic happening from the replacement of tropical forests by agricultural land is more relevant, as this increases the contact between humans and wild animals, either by animals sharing land with agriculture, or by simple displacement of species that migrate to areas already occupied by humans. Even simple urban expansion may trigger a zoonotic disease.

Essentially, the increased contact between humans and wild animals is an outcome of economic activities. These activities may be local, international or a combination of both. Although it is hard to trace the relationship between consumption in a Western city and population growth in a particular urban area in Asia, a more direct relation can be noted between consumption and deforestation, particularly in tropical forest areas (Lafferty, 2009; López and Galinato, 2005). For example, the consumption of timber in developed countries, such as the United States, Germany and Japan, is supported by developing nations in spite of the fact that these developed nations have considerable areas of forests (Zhang et al., 2020).

Agriculture is also a problem. The conversion of the Amazon forest to livestock farming is a leading cause of deforestation (Silva et al., 2020). In 2020, Brazil was the largest meat exporter in the world (Statista, 2022), with China and the European Union being the top destinations for Brazilian meat (ABIEC, 2020). In Germany, 12% of cattle meat comes from Argentina (TrendEconomy, 2020), while the Netherlands import meat from a staggering 80 different countries (Province of Manitoba, 2021), although curiously, it is also the EU's largest meat exporter (Statistics Netherlands, 2021). There is another important aspect of the Netherlands meat industry. To support local meat production, many products are imported. The local herds are dependent on maize, crude palm oil and soybeans from countries outside of the European Union (Statistics Netherlands, 2021). This is a common cycle. Often the pressure over forests, particularly in developing countries, happens in order to supply international demand. For that reason, according to Allen et al. (2017), the global hotspots of emerging zoonotic diseases are in tropical areas with high land-use changes and where wildlife biodiversity (mammal species richness) is high. The COVID-19 pandemic may not be an example of this cycle as the origin of the current pandemic remains unknown. The most possible scenario is that the virus came from a bat, transmitted directly to humans or through an intermediate animal (Maxmen and Mallapaty, 2021).

The international economic structure is responsible for the cycle of pressure on tropical areas, deforestation, displacement of wild animals and increased interaction between humans and wildlife. In addition, because international travel and trade are so common, easy and cheap, epidemic diseases become pandemic (Dharani et al., 2020). The 21st century has already witnessed many zoonotic diseases. For instance, swine flu spread in 58 countries and severe acute respiratory syndrome (SARS) occurred in 10 (Dharani et al., 2020). Ebola, perhaps the most lethal zoonotic disease, has occurred since 1976 in more than ten countries with lethality of around 50% (World Health Organization, 2022). Although not all zoonotic diseases are an outcome of the economic system, as cultural aspects and traditions may also cause contact between humans and wild species, this chapter focuses on the mechanism of international product consumption in urban areas and its role in deforestation, potentially contributing to pandemics.

It is important to understand the relationship between an economy and a pandemic. A deregulated or low-regulated economy is uninterested in understanding the entire production chain, and hence the possible connection to new pandemics. Meat produced in the Netherlands with imported feed exemplifies a process in which climate change, deforestation and pandemics are intrinsic parts, despite the home-produced meat seeming to be a local product. Although the European Union can impose severe restrictions on products that are not sustainable under its environmental requirements for importation, the production of these imported products may be based on a multi-webbed network of small suppliers in areas in which law enforcement is impractical, too expensive or simply impossible. In fact, some supermarkets in Europe have dropped Brazilian meat as the production was linked to deforestation following the outcome of a newspaper investigation alongside the Bureau of Investigative Journalism, which identified "cattle laundering, where cattle raised in illegally deforested areas are then brought to legal farms to be fattened and sent to slaughter (*The Guardian*, 2021b).

The consequence is that even a very well-regulated economy may struggle to guarantee that a certain production chain is not a pandemic risk, or perhaps that the production may be worth the risk. While assessment of a production chain with its hundreds of suppliers may be complex and expensive, the limitation of the consumption of products that are known for their international production network complexity is, in theory, simpler. However, limiting the final consumption may be perceived as a freedom restriction. In order to ensure a pandemic-free production chain, the assessment of its stages needs to be unbiased, auditable and should disregard economic impact. As soon as economic impact is included in the assessment, the possibility of a pandemic becomes a risk assessment exercise instead of a scientific identification of risk. While societies may not be willing to face a pandemic, the economy may be. Finally, it is important to identify the complexity of the risk assessment, as the risk of a virus reaching a human may be known but predictions regarding lethality and transmissibility are not. Fan et al. (2018) have shown that extensive assessments have been made regarding potential losses from pandemics in terms of income where the mortality loss is virtually ignored. Although terrible

TABLE 14.2 21st-century death toll from pandemics

Year	*Disease*	*Deaths*	*Source*
2003	Severe acute respiratory syndrome (SARS)	774	WHO (2003)
2007	Influenza A H1N5 (bird flu)	282	WHO/GIP (2015)
2009	Influenza H1N1 (swine flu)	201,200	Dawood et al. (2012)
2012	Middle East respiratory syndrome (MERS)	851	WHO—Eastern Mediterranean (2019)
2014	Ebola	11,325	Kaner and Schaack (2016)
2020	COVID-19 (up to 22 January 2022)	5,580,438	Worldometer (2022)

and possibly immoral, this is understandable in economic terms. The bottom-line assessment is not a record of what happened to achieve that result. It means that for the economy the result is more important than death from a pandemic.

A 2015 article by Ross et al. (five years before COVID-19), titled "Planning for the Next Global Pandemic," raised important factors regarding pandemic risks and pandemic frequency. First, pandemics are becoming more frequent. While pandemics happened roughly every decade in the 20th century, in the 21st century, they have become more common (Table 14.2). Overpopulation and poverty are the main contributing factors that have brought about this change and are strongly linked with climate change, environmental impact, natural habitat loss and increased human/host/reservoir interaction (Ross et al., 2015).

Beside the direct contact between humans and wildlife, there is evidence that the increase in temperature as a result of climate change may lead to an increase in transmission, for example, by a longer active season for mosquitoes or through new animal migration routes that lead to species not previously in contact starting to interact (Curseu et al., 2010).

The 1999 report *Livestock to 2020: The Next Food Revolution*, produced jointly by the International Food Policy Research Institute (IFPRI), the UN Food and Agriculture Organization (FAO) and the International Livestock Research Institute, summarised changes in the production of meat that have subsequently happened. According to the report, "the livestock revolution will stretch the capacity of existing production and distribution system and exacerbate environmental and public health problems" (Delgado et al., 1999). The authors pointed out the risk from zoonotic diseases that would be the outcome of the intensification of confined livestock fed by cereals. Although this report only addressed diseases arising from contact between a huge number of animals and humans, as noted above, another potential source of zoonotic disease is the pressure on forests.

Apart from forests being cut down and replaced with farmland, pressure on natural areas may come from industries such as logging or mineral extraction. Consequently, the risk of pandemics comes from the way people eat to the direct increase in CO_2 in the atmosphere increasing global temperature, suggesting that urban areas may be contributing to the development of new pandemics through the demands they make on land to supply their needs.

14.4 Large-scale pandemics and CO_2 reduction

The sheer scale of COVID-19 and its death toll has caused enormous disruption. In order to reduce deaths and to avoid collapse of health systems, authorities around the world restricted people's movements, and sometimes imposed lockdowns, which necessitated working-from-home policies. Although with differences, lockdowns around the world have had similar goals. During late March and early April 2020, all 19 cities had some sort of restriction (Table 14.3).

There were two main outcomes from such restrictions that affected not just the 19 cities but the world. The first, identified by scientists and non-scientists alike, was an improvement in air quality. While scientists were able to record a reduction in air pollution, citizens everywhere noticed better air quality and less smog. An assessment by Park et al. (2021) suggested a pollution reduction of around 40% in Seoul. Similar results were found by Wang et al. (2020) who showed that lockdown restrictions reduced energy related emissions by 18%, with 60% of this reduction coming from reduced transportation, followed by construction (23%) and other industries (12%).

These reductions are essentially the immediate outcome of the restrictions, although there is a more subtle but equally powerful effect of these restrictions—lack of consumption. As the lockdowns differed, the impact on consumption also differed. Some parts of the world were more affected than others depending on the level of restrictions and local markets. In some countries or cities where lockdowns were imposed and enforced, the possibilities for consumption were reduced to the basics of food and medicine, New Zealand being an example. It had a very restrictive lockdown in March 2020 that lasted six weeks with consumption being reduced to essential items determined by the government. It forced people to reset their consumption behaviour. Even online shopping was not possible during that period (you could buy overseas, but it would not be delivered). For six weeks, society had to reinvent itself in New Zealand (Unite Against Covid-19, 2020). To some extent, this also happened in other parts of the world. This exceptional behaviour suggests two very distinctive aspects of lockdown. First, people may be able to change quickly and dramatically if their lives seem to be in danger. Second, it was possible to see that people would be able to live with less and disengage from the consumption chain within which most of the Western nations live.

There were other COVID-induced behavioural changes that relate back to the environment apart from less consumption. For those who could work from home, and had appropriate space to do so, the use of cars or public transport was reduced, reducing fossil fuel emissions. In many places driving to distant shops was not allowed, and local consumption was necessary. It is expected that post-pandemic many people will remain working from home, or at least reducing hours in the office (Global Workplace Analytics, n.d.). This work-living arrangement is likely

TABLE 14.3 Types of restriction and dates imposed

City	*Restrictions*	*Date*	*Source*
Ahmedabad	Lockdown	07/05/2020	NDTV (2020)
Amsterdam	Lockdown	15/03/2020	DutchNews (2020)
Auckland	Lockdown	25/03/2020	Unite Against Covid-19 (2020)
Beijing	Lockdown	10/02/2020	*South China Morning Post* (2020)
Buenos Aires	Lockdown	20/03/2020	Bloomberg (2020)
Cairo	Lockdown	25/03/2020	Reuters (2020c)
Copenhagen	Danish authorities restricted public assembly to 10 or fewer people, ordered the closure of schools, universities, day cares, restaurants, cafes, libraries, gyms and hair salons, and shut all borders to most foreigners	11/03/2020	Reuters (2020b)
Dubai	Lockdown	24/03/2020	CNBC (2020)
Hamburg	Restaurants and gatherings of more than six people banned	20/03/2020	DW (2020)
Havana	Lockdown	03/04/2020	*Independent* (2020)
Jakarta	Partial Lockdown - offices, schools, places of worship and public spaces closed; transportation capacity within region limited but no authority to close inter-regional transportation services	03/04/2020	*The Jakarta Post* (2020)
Kuala Lumpur	Lockdown	18/03/2020	*Forbes India* (2020)
Moscow	Lockdown	30/03/2020	*The Guardian* (2020a)
Nairobi	Lockdown	06/04/2020	Al Jazeera (2020)
New York	Lockdown	18/03/2020	BBC (2020)
Pyongyang*	Borders closed	March 2020	BBC News (2020); Business Insider (2020); *First Post* (2020)
São Paulo	Lockdown	24/03/2020	Reuters (2020a)
Singapore	Lockdown	04/04/2020	*Mint* (2020); *The Guardian* (2020c)
Sydney	Lockdown	31/03/2020	NSW Government (2020)

* Information regarding internal restrictions in North Korea not found.

to affect the suburban/city centre relationship, reducing the demand for office buildings and even decreasing the demand for public transport.

The potential disruption to consumption from the 2020 pandemic may be one of the reasons many leaders around the world have denied the dangers of COVID-19. For example, Brazilian president Jair Bolsonaro, besides saying, on 24 March 2020 that no more than 800 people would die from COVID in Brazil (BBC Brasil, 2020; *Metropoles*, 2020) and COVID-19 was no more than a "little cold" (BBC Brasil, 2020), has maintained on many occasions that the economy is more important than COVID and that lockdowns would cause more harm than the pandemic (CNN Brasil, 2021). His counterpart in the United States during the first year of the pandemic, Donald Trump, had similar attitudes. In May 2020, in the early days of the pandemic, *The Washington Post* compiled the 54 times that Trump had downplayed the pandemic. Statements like "views this the same way as the flu" and "it will disappear as a miracle" were said on many occasions (*The Washington Post*, 2020). On 24 March, Trump also said "we gonna lose more people by putting a country to a massive recession or depression" (*The Washington Post*, 2020). Curiously, this was the same day that Bolsonaro mentioned no more than 800 people would die. As of, January 2022, according to Worldometer (2022), Brazil and the United States have, combined, more than 37% of the total deaths in the world due to COVID-19 (Brazil 623,145 and the United States 890,052 deaths). The figures will be higher by the time this book is published.

Both presidents are similarly aligned in denying climate change. Donald Trump, who withdrew from the Paris Agreement in 2017 (*The New York Times*, 2017), has challenged climate science (Aldy, 2017). Brazil's former foreign minister during Bolsonaro's presidency, Ernesto Araújo, stated climate change was a conspiracy organised by "cultural Marxists" (*The Guardian*, 2020b). Bolsonaro, in December 2019, claimed global warming was a commercial game (UOL Meio Ambiente, 2019).

For both the pandemic and climate change, the mechanism of denial is essentially the same (see Chapter 3). This process aims to oppose scientific facts with political opinion. The idea is to cause enough doubt to encourage people to continue to behave in the same way and keep the economy going. It is a strategy created to maintain the status quo, but one that is cruel, whether for climate change or the pandemic, as it may put people's lives at risk. Even in the face of overwhelming evidence and scientific support proving climate change and the risks of COVID-19, a small minority of well-presented, dissenting scientists or public figures delivering information supported by doubtful science are enough to convince some people. This process is well documented, particularly with climate change, and has been effective. It plays with the psychology of science denial, which is the refusal to believe in existing scientific facts (Björnberg et al., 2017). Externally motivated denial is well researched for climate change, which is a direct target where the desire to avoid scientific facts is achieved by biased cognitive strategy (Wong-Parodi and Feygina, 2020). It is simple, the pandemic and climate change threaten the status quo, and people naturally gravitate towards keeping things as they are (Wong-Parodi and Feygina, 2020). In addition, the organised dissemination of misinformation is directed to those who are considered most likely to yield the desired conclusion (Kunda, 1990). As a result, any speech that supports the status quo will find an audience among people, even if it is against science.

There is a psychological opportunity for people to support the current economic system by presenting information that they want to believe, whether this is calling climate change a hoax or COVID-19 the flu. This is, perhaps, the main reason for the global protests against lockdowns and vaccination although such claims are unsupported by serious science. Table 14.4 gives examples of protests in the 19 cities.

There is little doubt that many protesters based their opinions on misinformation and reluctance to accept science. This process of challenging science, minimising the risk of COVID-19 and questioning the efficacy of vaccines, makes the fight

TABLE 14.4 Examples of protests from 12 of the 19 cities

Ahmedabad	NGO calls for protests in multiple cities across India against anti-COVID measures as similar protests are being carried out across the world	Opindia (2021)
Amsterdam	Dutch police disperse thousands protesting in Amsterdam against COVID lockdown	France 24 (2022)
Auckland	Thousands at Auckland lockdown protest	*New Zealand Herald* (2021)
Buenos Aires	Argentina receives AstraZeneca jabs amid anti-lockdown protests	Al Jazeera (2021)
Copenhagen	Hundreds of Danes protest against COVID restrictions in Copenhagen	Euronews (2021a)
Hamburg	Thousands of people have demonstrated against coronavirus rules in several major German cities, including Hamburg and Düsseldorf. Pandemic deniers have become increasingly displeased with virus-related regulations	DW (2022)
Jakarta	Further protests against extension of COVID-19 restrictions likely nationwide through late July	Crisis 24 (2021)
Moscow	Russian Vaccine sceptics oppose COVID passes	*The Moscow Times* (2021)
Nairobi	Kenya protesters clash with police at lockdown demo	Euronews (2021b)
New York	Hundreds of Anti-mandate protesters march as NYC workers' deadline looms	NBC New York, (2022)
São Paulo	Protesters are against lockdown restrictions in São Paulo	R7 (2021)
Sydney	Sydney anti-lockdown protests: violent clashes as police arrest demonstrators	*The Guardian* (2021a)

against COVID-19 even harder. Despite efforts to challenge science and thereby keep the economic system intact, there has been a good uptake of vaccines and despite lockdown fatigue, some people reduced their normal movements, which potentially harms the economy, in order to stay safe (Watanabe and Yabu, 2021). The economy is not a force that interferes in public health decisions or issues by itself, it has and needs political support. Discussion of the need for lockdowns and vaccines hurts the very core principle of neoliberalism, which brings with it a connection between the pandemic and the environment. Apparently limiting is a concept not fully accepted by modern society. For part of society, limiting goes against freedom and these concepts find strong support in the economic principle of neoliberalism. According to Harvey (2006), neoliberalism is a political economic practice, which proposes that societies will advance following the principle of maximisation of entrepreneurial and individual freedom. Harvey goes on to point out that neoliberalism advocates against state intervention. Therefore, lockdown and limiting the movement of those who choose not to be vaccinated may be perceived as an attack on democracy itself. Giannone (2010) adds that neoliberalism suggests a version of democracy where freedom is more important than other democratic values such as equality. Gutmann (1993) suggested that reconciling freedom and civic virtue was an educational challenge. Our current social situation, in relation to both the pandemic and the environment, is an example of the problem identified by Gutmann. For the pandemic, vaccination is more than an individual decision. A vast majority of people need to be vaccinated to halt the circulation of the virus, making vaccination both an individual and collective effort (European Vaccination Information Portal, 2022; WHO, 2020a). It is essentially the same with the environment; society will only perceive a difference if enough people make considerable changes, and this may go against the economic system of today.

Although not all the cities in this book are organised under a neoliberal economic structure, it is useful to look at the discussion between individual freedom and collective interest, as it reflects some of the neoliberal principles. It is fair to assume that the neoliberal idea has strong roots among societies, even in countries that do not operate following all neoliberal principles. For discussions that involve the environment or the pandemic, it is convenient, under the economic point of view, to keep the idea that democracy means total individual freedom, which is not, and has never been, true. For instance, the text of the European Convention for the Protection of Human Rights and Fundamental Freedoms states that these freedoms may be subject to restrictions or penalties prescribed by law and are necessary in a democratic society in the interest of public safety (Council of Europe, 1950). Such constraints have been more visible during the pandemic where limits to movement have been imposed in several parts of the world (Table 14.3).

Although there has been a constant effort to discredit science and the dangers of COVID-19, many people have taken it seriously and followed the recommendation of authorities based on science. It is intriguing but not surprising that climate change has not triggered a similar reaction and behaviour change. One of the reasons cited in academic literature is the psychological distance of climate change.

According to Spence et al. (2012), climate change is perceived as a distant issue. In fact, changing environmental behaviour is difficult even when the risks appear closer (Brügger et al., 2015; Chu and Yang, 2018; Scannell and Gifford, 2016). It seems people perceive the environmental issues related to climate change as something temporally, geographically and socially distant. The fourth factor that may influence people not to act more decisively towards the environment is uncertainty. Scientists cannot precisely state what will happen in the future regarding climate change (Spence et al., 2012).

The confusion between individual freedom, total freedom and democracy plays an important role in gathering people to fight against science. The understanding that individual freedom does not mean total freedom and the collective interest is also important is conveniently not clear, encouraging people to challenge scientific facts and disengage from behavioural change. Protests and demonstrations are still happening with people challenging vaccination mandates. The discussion here is not if people should or should not be obligated to vaccinate, the point of concern is the reason behind the hesitancy to be vaccinated. Dubé et al. (2021) claimed that rumours that the pandemic is a hoax to sell vaccines were spreading faster than the virus on social media, which led to the increase in vaccination hesitancy. There have been studies showing that non-vaccinated people are more likely to develop serious illness (Havers et al., 2021; Tenforde et al., 2021), but people have still challenged these data. Whatever the reasons discouraging people to be vaccinated, some continue to believe that their democratic right entitles them to not take the vaccine and not have to face any restrictions because this is their choice.

The concepts of and relationships between democracy and individual freedom are not simple. Democracy is the process of providing the people of a region or country with a say in decisions on all matters that may affect them (Radder, 2021). In many countries, the democratic process allows lobbies that may manipulate public and government opinion (see Chapter 9). Individual freedom, however, is a concept that is in many aspects not present in modern society. Individuals are bounded by laws and social norms, nonetheless, a democracy should entitle freedom of speech (Radder, 2021). The theoretical and universal freedom of speech allows individuals to promote bad science, to deny climate change and to advocate against vaccination. Freedom of speech now goes beyond an individual speaking to a group of people. Through social media and the internet, the power of freedom of speech has been catalysed. Most of the social media companies, all of which exist solely to make a profit, allow paid promotions, meaning any stakeholder can reach millions of people, broadcasting any information they want (see Chapter 12). This could be viewed as the combination of freedom of speech and capitalism. This is another point of convergence between the pandemic and climate change denial. Essentially, self-promotion in social media is not the main concern of scientists, thus it becomes a competition between scientists' disregard for self-promotion on social media and a group of people or companies using the promotion capacity of social media to promote their agenda. As opposed to traditional media, social media has little or no filter, no fact-checking nor editorial judgement (Allcott and Gentzkow, 2017).

14.5 Conclusion

This chapter has explored common aspects between climate change, environmental impact and COVID-19, beginning with identifying that economic growth plays an important role in deforestation, which in turn increases the risk of new pandemics. The global economic system that connects cities in the supply and demand relationship, is so vast and complex that it is almost impossible to ensure that all production chains are not contributing directly or indirectly to new pandemics. International trade and travel assist the quick dissemination of viruses and zoonotic diseases.

There are similarities between people who deny climate change and those who deny facts related to the pandemic. In both cases, the goal is to keep the economy moving without any interruption. It is worrying at the same time that the capacity to reach out to people with misinformation has increased dramatically and trends upwards.

There are, however, some items of good news. In the first months of the pandemic in 2020, when countries were establishing the first lockdowns, there was a massive change in people's behaviour, proving that people and society are capable of rapid change. In January 2020 it was unimaginable to think it would not be possible to consume goods, travel or socialise for weeks. This economic shift reduced CO_2 emissions, both locally and internationally (Aktar et al., 2021; Wang et al., 2020b; Zeng et al., 2021). Unfortunately, after the lockdowns, CO_2 emissions returned to similar levels prior to the pandemic and the stimulus in post-pandemic economic packages may even increase the levels again (Schulte-Fischedick et al., 2021). It is also expected that after lockdowns those who keep their jobs will have disposable money available, which may lead to higher consumption. However, the main point is that as a society, we can adjust and if the issues related to the environment and climate change are properly communicated, we may see more effective changes. People just need to understand that we are facing an environmental challenge as serious as the pandemic, if not more so.

References

ABIEC (2020). Beef Report 2020. Retrieved from http://abiec.com.br/publicacoes/beef-report-2020/#dflip-df_2947/1/, 5 January 2022.

Aktar, M.A., Alam, M.M. and Al-Amin, A.Q. (2021). Global Economic Crisis, Energy Use, CO_2 Emissions, and Policy Roadmap Amid COVID-19. *Sustainable Production and Consumption 26*, 770–781.

Al Jazeera (2020). COVID-19: Kenya Bans Travel in and Out of Nairobi, Other Areas. Retrieved from www.aljazeera.com/news/2020/4/6/covid-19-kenya-bans-travel-in-and-out-of-nairobi-other-areas, 10 January 2022.

Al Jazeera. (2021). *A*rgentina Receives AstraZeneca Jabs Amid Anti-lockdown Protests. Retrieved from www.aljazeera.com/news/2021/4/18/argentina-receives-astrazeneca-jabs-amid-anti-lockdown-protests, 15 January 2022.

Aldy, J.E. (2017). Real World Headwinds for Trump Climate Change Policy. *Bulletin of the Atomic Scientists 73*(6), 376–381.

Allcott, H. and Gentzkow, M. (2017). Social Media and Fake News in the 2016 Election. *Journal of Economic Perspectives*, *31*(2), 211–235.

Allen, T., Murray, K.A., Zambrana-Torrelio, C., Morse, S.S., Rondinini, C., di Marco, M., Breit, N., Olival, K.J. and Daszak, P. (2017). Global Hotspots and Correlates of Emerging Zoonotic Diseases. *Nature Communications 8*(1), 1–10.

Andrée, B.P.J., Chamorro, A., Spencer, P., Koomen, E. and Dogo, H. (2019). Revisiting the Relation Between Economic Growth and the Environment: A Global Assessment of Deforestation, Pollution and Carbon Emission. *Renewable and Sustainable Energy Reviews*, *114*, 16pp.

BBC (2020). New York: The City that Never Sleeps on Lockdown. Retrieved from www.bbc.com/news/business-51880799, 14 January 2022.

BBC Brasil. (2020). 2 momentos em que Bolsonaro chamou covid-19 de "gripezinha", o que agora nega. Retrieved from www.bbc.com/portuguese/brasil-55107536, 14 January 2022.

BBC News. (2020). Coronavirus: North Korea Quarantines Foreigners. Retrieved from www.bbc.com/news/world-asia-51609360, 14 January 2022.

Björnberg, K.E., Karlsson, M., Gilek, M. and Hansson, S.O. (2017). Climate and Environmental Science Denial: A Review of the Scientific Literature Published in 1990–2015. *Journal of Cleaner Production*, *167*, 229–241.

Bloomberg (2020). Argentina Orders "Exceptional" Lockdown in Bid to Stem Virus. Retrieved from www.bloomberg.com/news/articles/2020-03-20/argentina-orders-exceptional-lockdown-in-bid-to-contain-virus, 14 January 2022.

Brancalion, P.H.S., Broadbent, E.N., de-Miguel, S., Cardil, A., Rosa, M.R., Almeida, C.T., Almeida, D.R.A., Chakravarty, S., Zhou, M., Gamarra, J.G.P., Liang, J., Crouzeilles, R., Hérault, B., Aragão, L.E.O.C., Silva, C.A. and Almeyda-Zambrano, A.M. (2020). Emerging Threats Linking Tropical Deforestation and the COVID-19 Pandemic. *Perspectives in Ecology and Conservation*, *18*(4), 243–246.

Brügger, A., Dessai, S., Devine-Wright, P., Morton, T. A., and Pidgeon, N. F. (2015). Psychological responses to the proximity of climate change. *Nature Climate Change 2015 5:12*, *5*(12), 1031–1037.

Business Insider (2020). 23 Photos Show What Life in North Korea Is Like during Its Coronavirus Lockdown. Retrieved from www.businessinsider.com.au/north-korea-coronavirus-lockdown-photos-kim-jong-un-2020-4?r=US&IR=T, 14 January 2022.

Chu, H., and Yang, J. Z. (2018). Taking climate change here and now – mitigating ideological polarization with psychological distance. *Global Environmental Change*, *53*, 174–181.

CNBC (2020). Coronavirus Lockdown: Dubai Requires Police Permits for Leaving the House. Retrieved from www.cnbc.com/2020/04/07/coronavirus-lockdown-dubai-requires-police-permits-for-leaving-the-house.html, 14 January 2022.

CNN Brasil (2021). Covid-19: Bolsonaro volta a criticar medidas restritivas no Brasil. Retrieved from www.cnnbrasil.com.br/politica/covid-19-bolsonaro-volta-a-criticar-medidas-restritivas-no-brasil/, 14 January 2022.

Council of Europe. (1950). Details of Treaty No.005. Retrieved from www.coe.int/en/web/conventions/full-list?module=treaty-detail&treatynum=005, 15 May 2022.

Crespo Cuaresma, J., Danylo, O., Fritz, S., Mccallum, I., Obersteiner, M., See, L. and Walsh, B. (2016). Economic Development and Forest Cover: Evidence from Satellite Data. *Scientific Reports* 7(1), 8pp.

Crisis 24 (2021). Indonesia: Further Protests Against Extension of COVID-19 Restrictions Likely Nationwide Through Late July. Retrieved from https://crisis24.garda.com/insights-intelligence/intelligence/risk-alerts/wip1001304531/indonesia-further-protests-against-extension-of-covid-19-restrictions-likely-nationwide-through-late-july, 14 January 2022.

Curseu, D., Popa, M., Sirbu, D. and Stoian, I. (2010). Potential Impact of Climate Change on Pandemic Influenza Risk. *Global Warming, 31*, 643–657.

Dawood, F.S., Iuliano, A.D., Reed, C., Meltzer, M.I., Shay, D.K., Cheng, P.Y., Bandaranayake, D., Breiman, R.F., Brooks, W.A., Buchy, P., Feikin, D.R., Fowler, K.B., Gordon, A., Hien, N.T., Horby, P., Huang, Q.S., Katz, M.A., Krishnan, A., Lal, R... Widdowson, M.A. (2012). Estimated Global Mortality Associated with the First 12 Months of 2009 Pandemic Influenza A H1N1 Virus Circulation: A Modelling Study. *The Lancet Infectious Diseases 12*(9), 687–695.

Delgado, C., Rosengrant, M., Steinfeld, H., Ehui, S. and Courbois, C. (1999). Livestock to 2020 The Next Food Revolution. Retrieved from www.fao.org/Ag/againfo/resources/documents/lvst2020/20201.pdf, 10 January 2022.

Dharani, K.G., Mohanarathinam, A., Mansoor, S., Daranikumar, M. and Sasikala, P. (2020). Deforestation is a Cause of Pandemic—A Systematic Analysis. *Proceedings of the 4th International Conference on Electronics, Communication and Aerospace Technology, ICECA 2020*. 1491–1497.

Dubé, È., Ward, J.K., Verger, P. and Macdonald, N.E. (2021). Vaccine Hesitancy, Acceptance, and Anti-Vaccination: Trends and Future Prospects for Public Health. *Annual Review of Public Health 42*(1), 175–191.

DutchNews (2020). The Netherlands on Lockdown: Schools, Cafes and Sports Clubs Shut. Retrieved from www.dutchnews.nl/news/2020/03/the-netherlands-on-lockdown-schools-cafes-and-sports-clubs-shut/, 14 January 2022.

DW (2020). German States Move Closer to Near-total Lockdowns. Retrieved from www.dw.com/en/german-states-move-closer-to-near-total-lockdowns/a-52863482, 14 January 2022.

DW (2022). Germany: Thousands Protest Against COVID Measures Across Country. Retrieved from www.dw.com/en/germany-thousands-protest-against-covid-measures-across-country/a-60181181, 14 January 2022.

Euronews (2021a). Hundreds of Danes Protest Against COVID Restrictions in Copenhagen. Retrieved from www.euronews.com/2021/02/28/hundreds-of-danes-protest-against-covid-restrictions-in-copenhagen, 14 January 2022.

Euronews (2021b). Kenya Protesters Clash with Police at Lockdown Demo. Retrieved from www.euronews.com/2021/07/07/kenya-protesters-clash-with-police-at-lockdown-demo, 14 January 2022.

European Vaccination Information Portal (2022). Benefits of Vaccination for the Community. Retrieved from https://vaccination-info.eu/en/vaccination/benefits-vaccination-community, 18 January 2022.

Fan, V.Y., Jamison, D.T. and Summers, L.H. (2018). Pandemic Risk: How Large Are the Expected Losses? *Bulletin of the World Health Organization 96*(2). 129.

First Post (2020). North Korea Flies Out Dozens of Diplomats, Foreigners to Russia before Pyongyang Tightens Lockdown to Contain Coronavirus Outbreak. Retrieved from www.firstpost.com/health/north-korea-flies-out-dozens-of-diplomats-foreigners-to-russia-before-pyongyang-tightens-lockdown-to-contain-coronavirus-outbreak-8133101.html, 14 January 2022.

Forbes India (2020). What the Covid-19 Lockdown Looks Like in...Kuala Lumpur, Malaysia. Retrieved from www.forbesindia.com/article/coronavirus/what-the-covid19-lockdown-looks-like-inkuala-lumpur-malaysia/58949/1, 14 January 2022.

France 24 (2022). Dutch Police Disperse Thousands Protesting in Amsterdam against Covid Lockdown. Retrieved from www.france24.com/en/europe/20220102-dutch-police-disperse-thousands-protesting-in-amsterdam-against-covid-lockdown, 18 January 2022.

Giannone, D. (2010). Political and Ideological Aspects in the Measurement of Democracy: The Freedom House Case. *Democratization 17*(1). 68–97.

Global Workplace Analytics (n.d.). Work-at-Home after Covid-19: Our Forecast. Retrieved from https://globalworkplaceanalytics.com/work-at-home-after-covid-19-our-forecast, 15 May 2022.

The Guardian (2020a). "It's like a normal day": Moscow Grapples with Virus Lockdown. Retrieved from www.theguardian.com/world/2020/mar/30/moscow-coronavirus-lockdown-stay-indoors-city, 18 January 2022.

The Guardian (2020b). Resistance to the "Environmental Sect" Is a Cornerstone of Bolsonaro's Rule. Retrieved from www.theguardian.com/global-development/2020/jul/27/resistance-to-the-environmental-sect-is-a-cornerstone-of-bolsonaro-rule-brazil, 18 January 2022.

The Guardian (2020c). Singapore Extends Lockdown as Covid Cases Surge Past 9,000. Retrieved from www.theguardian.com/world/2020/apr/21/singapore-coronavirus-outbreak-surges-with-3000-new-cases-in-three-days, 18 January 2022.

The Guardian (2021a). Melbourne and Sydney Anti-lockdown Protests: Violent Clashes as Police Arrest Demonstrators. Retrieved from www.theguardian.com/australia-news/2021/sep/18/victoria-and-nsw-lockdown-protests-violent-clashes-as-police-arrest-demonstrators, 18 January 2022.

The Guardian (2021b). Supermarkets Drop Brazilian Beef Products Linked to Deforestation. Retrieved from www.theguardian.com/environment/2021/dec/16/supermarkets-drop-brazilian-beef-products-linked-to-deforestation?s=03, 12 January 2022.

Gutmann, A. (1993). Democracy and Democratic Education. *Studies in Philosophy and Education 12*, 1–9.

Harvey, D. (2006). Neo-Liberalism as Creative Destruction. *Geografiska Annaler: Series B, Human Geography 88*(2), 145–158.

Havers, F.P., Pham, H., Taylor, C.A., Whitaker, M., Patel, K., Anglin, O., Kambhampati, A.K., Milucky, J., Zell, E., Chai, S.J., Kirley, P.D., Alden, N.B., Armistead, I., Yousey-Hindes, K., Meek, J., Openo, K.P., Anderson, E.J., Reeg, L., Kohrman, A… McMorrow, M. (2021). COVID-19-associated Hospitalizations among Vaccinated and Unvaccinated Adults ≥18 Years—COVID-NET, 13 states, 1 January 1–24 July 2021. *Medical Letter on the CD and FDA*. 91.

Independent (2020). Cuba Bans All Tourists and Imposes Draconian Rules amid Coronavirus Pandemic. Retrieved from www.independent.co.uk/travel/news-and-advice/coronavirus-cuba-tourists-ban-travel-lockdown-havana-quarantine-a9446466.html, 15 January 2022.

The Jakarta Post (2020). 50 Days of Indonesia's Partial Lockdown. Is it Enough for the "New Normal"? Retrieved from www.thejakartapost.com/news/2020/05/28/50-days-of-indonesias-partial-lockdown-is-it-enough-for-the-new-normal.html, 18 January 2022.

Kaner, J. and Schaack, S. (2016). Understanding Ebola: The 2014 Epidemic. *Globalization and Health. 12*(1), 7pp.

Kunda, Z. (1990). The Case for Motivated Reasoning. *Psychological Bulletin 108*(3), 480–498.

Lafferty, K.D. (2009). The Ecology of Climate Change and Infectious Diseases. *Ecology 90*(4), 888–900.

López, R. and Galinato, G.I. (2005). Trade Policies, Economic Growth, and the Direct Causes of Deforestation. *Land Economics 81*(2), 145–169.

Madhav, N., Oppenheim, B., Gallivan, M., Mulembakani, P., Rubin, E. and Wolfe, N. (2017). Pandemics: Risks, Impacts, and Mitigation. In Jamison, D.T., Gelband, H., Horton, S., Jha, P., Mock, C.N. and Nugent, R. (eds). *Disease Control Priorities: Improving Health and Reducing Poverty*. Washington, DC: The World Bank Group. 315–345.

Maxmen, A. and Mallapaty, S. (2021). The COVID Lab-leak Hypothesis: What Scientists Do and Don't Know. *Nature*, *594*(7863), 313–315.

Metropoles (2020). Covid-19: na crise, Bolsonaro prioriza discurso econômico. Retrieved from www.metropoles.com/brasil/jair-bolsonaro/covid-19-na-crise-bolsonaro-prioriza-discurso-economico, 13 January 2022.

Mint (2020). Singapore Announces One-month Lockdown. Retrieved from www.livemint.com/news/world/singapore-announces-1-month-lockdown-amid-rising-corona-cases-11585905344421.html, 14 January 2022.

The Moscow Times (2021). Russian Vaccine Skeptics Oppose Covid Passes. Retrieved from www.themoscowtimes.com/2021/11/29/russian-vaccine-skeptics-oppose-covid-passes-a75695, 19 January 2022.

NBC New York (2022). NYC Vaccine Mandate Looms amid Protests. Retrieved from www.nbcnewyork.com/news/local/more-anti-mandate-protests-expected-as-deadline-for-nyc-workers-looms/3348726/, 18 January 2022.

The New York Times (2017). Trump Will Withdraw U.S. From Paris Climate Agreement. Retrieved from www.nytimes.com/2017/06/01/climate/trump-paris-climate-agreement.html, 12 January 2022.

The New York Times (2022). Spurred by Omicron, Europe Sets Covid Infection Records Every Day. Retrieved from www.nytimes.com/2021/12/29/world/europe/europe-omicron-infection-record-covid.html, 14 January 2022.

NDTV (2020). Ahmedabad Complete Lockdown: Ahmedabad Shut Down for a Week, Surat from Saturday. Retrieved from www.ndtv.com/india-news/coronavirus-ahmedabad-shut-for-a-week-paramilitary-forces-patrol-streets-2224908, 18 January 2022.

New Zealand Herald (2021). Covid 19 Delta: Thousands at Auckland Lockdown Protest, Brian Tamaki Speaks, Police Keep Distance. Retrieved from www.nzherald.co.nz/nz/covid-19-delta-thousands-at-auckland-lockdown-protest-brian-tamaki-speaks-police-keep-distance/NYPJ6ETBX2UA6I5ZXBO5EKEFDI/, 20 January 2022.

NSW Government (2020). Supporting NSW during COVID-19. Retrieved from www.service.nsw.gov.au/covid-19, 20 January 2022.

Opindia (2021). NGO Calls for Protests in Cities Across India against Anti-Covid Measures. Retrieved from www.opindia.com/2021/07/ngo-calls-for-protests-in-cities-across-india-against-anti-covid-measures/, 18 January 2022.

Park, C., Jeong, S., Shin, Y, Cha, Y. and Lee, H. (2021). Reduction in Urban Atmospheric CO_2 Enhancement in Seoul, South Korea, Resulting from Social Distancing Policies during the COVID-19 Pandemic. *Atmospheric Pollution Research*. *12*(9), 7pp.

Province of Manitoba (2021). Beef Market in the Netherlands. Retrieved from www.gov.mb.ca/agriculture/markets-and-statistics/trade-statistics/pubs/beef-market-in-the-netherlands.pdf, 12 January 2022.

R7 (2021). Grupo faz protesto contra fase emergencial na zona sul de SP. Retrieved from https://noticias.r7.com/sao-paulo/grupo-faz-protesto-contra-fase-emergencial-na-zona-sul-de-sp-12032021, 18 January 2022.

Radder, H. (2021). Which Science, Which Democracy, and Which Freedom? In Hartl, P. and Tuboly, A.T. (eds). *Science, Freedom, Democracy*. New York: Routledge. 315–346.

Reuters (2020a). Biggest Brazil State's Governor Extends Lockdown through May 10. Retrieved from www.reuters.com/article/health-coronavirus-brazil-sao-paulo-idUSL1N2C50WH, 18 January 2022.

Reuters (2020b). Denmark Eyes Gradual Reopening After Easter if Coronavirus Numbers Stabilize. Retrieved from www.reuters.com/article/us-health-coronavirus-denmark-idUSKBN21H2LY, 18 January 2022.

Reuters (2020c). Egypt Declares Two-week Curfew to Counter Coronavirus. Retrieved from www.reuters.com/article/us-health-coronavirus-egypt-idUSKBN21B1MR, 18 January 2022.

Reuters (2022). Denmark to Offer Fourth Coronavirus Jab While Easing Curbs. Retrieved from www.reuters.com/world/europe/denmark-ease-coronavirus-restrictions-despite-omicron-surge-2022-01-12/, 20 January 2022.

Rohr, J.R., Barrett, C.B., Civitello, D.J., Craft, M.E., Delius, B., DeLeo, G.A., Hudson, P.J., Jouanard, N., Nguyen, K.H., Ostfeld, R.S., Remais, J.V., Riveau, G., Sokolow, S.H. and Tilman, D. (2019). Emerging Human Infectious Diseases and the Links to Global Food Production. *Nature Sustainability 2*(6), 445–456.

Ross, A.G.P., Crowe, S.M., Tyndall, M.W. and Petersen, E. (2015). Planning for the Next Global Pandemic. *International Journal of Infectious Diseases 38*, 89–94.

Rulli, M.C., Santini, M., Hayman, D.T.S. and D'Odorico, P. (2017). The Nexus between Forest Fragmentation in Africa and Ebola Virus Disease Outbreaks. *Scientific Reports* 7(1), 1–8.

Scannell, L., and Gifford, R. (2016). Place Attachment Enhances Psychological Need Satisfaction: *Environment and Behavior, 49*(4), 359–389.

Schulte-Fischedick, M., Shan, Y. and Hubacek, K. (2021). Implications of COVID-19 Lockdowns on Surface Passenger Mobility and Related CO_2 Emission Changes in Europe. *Applied Energy 300*, 11pp.

Silva, M.J.B., da Costa, M.F., de Farias, S.A. and Wanderley, L.S.O. (2020). Who Is Going to Save the Brazilian Amazon Forest? Reflections on Deforestation, Wildlife Eviction, and Stewardship Behavior. *Psychology & Marketing 37*(12), 1720–1730.

South China Morning Post (2020). Coronavirus Prompts Beijing Residential Lockdown as Millions Return to Work. Retrieved from www.scmp.com/economy/china-economy/article/3049847/coronavirus-prompts-beijing-residential-lockdown-millions, 18 January 2022.

Spence, A., Poortinga, W. and Pidgeon, N. (2012). The Psychological Distance of Climate Change. *Risk Analysis, 32*(6), 957–972.

Statista (2022). Global Major Beef Exporters in 2020. Retrieved from www.statista.com/statistics/917207/top-exporters-of-beef-globa/l, 12 January 2022.

Statistics Netherlands (2021). The Netherlands Is the EU's Largest Meat Exporter. Retrieved from www.cbs.nl/en-gb/news/2021/25/the-netherlands-is-the-eu-s-largest-meat-exporter, 12 January 2022.

Tenforde, M.W., Patel, M.M., Ginde, A.A., Douin, D.J., Talbot, H.K., Casey, J.D., Mohr, N.M., Zepeski, A., Gaglani, M., McNeal, T., Ghamande, S., Shapiro, N.I., Gibbs, K.W., Files, D.C., Hager, D.N., Shehu, A., Prekker, M.E., Erickson, H.L., Exline, M.C…Self, W.H. (2021). Effectiveness of Severe Acute Respiratory Syndrome Coronavirus 2 Messenger RNA Vaccines for Preventing Coronavirus Disease 2019 Hospitalizations in the United States. *Clinical Infectious Diseases 74*(9), 1515–1524.

TrendEconomy (2020). Germany | Imports and Exports | World | Meat of bovine animals | Value (US$) and Value Growth, YoY (%) | 2009–2020. Retrieved from https://trendeconomy.com/data/h2/Germany/0201, 10 January 2022.

Unite Against Covid-19 (2020). History of the COVID-19 Alert System. Retrieved from https://covid19.govt.nz/about-our-covid-19-response/history-of-the-covid-19-alert-system/, 15 January 2022.

UOL Meio Ambiente (2019). Bolsonaro diz que pressão sobre mudança climática é "jogo comercial". Retrieved from https://noticias.uol.com.br/meio-ambiente/ultimas-noticias/redacao/2019/12/15/bolsonaro-diz-que-pressao-sobre-mudanca-climatica-e-jogo-comercial.htm, 15 Januray 2022.

Wang, Q., Lu, M., Bai, Z. and Wang, K. (2020). Coronavirus Pandemic Reduced China's CO_2 Emissions in Short-term, While Stimulus Packages May Lead to Emissions Growth in Medium- and Long-term. *Applied Energy*, *278*, 7pp.

The Washington Post (2020). 54 Times Trump Downplayed the Coronavirus. Retrieved from www.washingtonpost.com/video/politics/the-fix/54-times-trump-downplayed-the-coronavirus/2020/03/05/790f5afb-4dda-48bf-abe1-b7d152d5138c_video.html, 15 January 2022.

Watanabe, T. and Yabu, T. (2021). Japan's Voluntary Lockdown. *PloS One 16*(6), 20pp.

WHO (2003). Summary of Probable SARS Cases with Onset of Illness from 1 November 2002 to 31 July 2003. Retrieved from www.who.int/publications/m/item/summary-of-probable-sars-cases-with-onset-of-illness-from-1-november-2002-to-31-july-2003, 14 January 2022.

WHO (2015). Cumulative Number of Confirmed Human Cases for Avian Influenza A(H5N1). Retrieved from www.who.int/influenza/human_animal_interface/EN_GIP_201503031cumulativeNumberH5N1cases.pdf, 12 January 2022.

WHO (2019). MERS Situation Update. Retrieved from https://applications.emro.who.int/docs/EMROPub-MERS-SEP-2019-EN.pdf?ua=1&ua=1, 18 January 2022.

WHO (2020a). Immunization Agenda 2030: A Global Strategy to Leave No One Behind. Retrieved from www.who.int/publications/m/item/immunization-agenda-2030-a-global-strategy-to-leave-no-one-behind, 18 January 2022.

WHO (2020b). WHO Director-General's Opening Remarks at the Media Briefing on COVID-19. Retrieved from www.who.int/director-general/speeches/detail/who-director-general-s-opening-remarks-at-the-media-briefing-on-covid-19-11-march-2020, 14 January 2022.

WHO (2022). Ebola Virus Disease. Retrieved from www.who.int/news-room/fact-sheets/detail/ebola-virus-disease, 14 January 2022.

Wong-Parodi, G. and Feygina, I. (2020). Understanding and Countering the Motivated Roots of Climate Change Denial. *Current Opinion in Environmental Sustainability 42*, 60–64.

Worldometer (2022). COVID Live—Coronavirus Statistics. Retrieved from www.worldometers.info/coronavirus/, 19 January 2022.

Zeng, N., Han, P., Liu, Z., Liu, D., Oda, T., Martin, C., Liu, Z., Yao, B., Sun, W., Wang, P., Cai, Q., Dickerson, R. and Maksyutov, S. (2021). Global to Local Impacts on Atmospheric CO_2 from the COVID-19 Lockdown, Biosphere and Weather Variabilities. *Environmental Research Letters 17*(1), 14pp.

Zhang, Q., Li, Y., Yu, C., Qi, J., Yang, C., Cheng, B. and Liang, S. (2020). Global Timber Harvest Footprints of Nations and Virtual Timber Trade Flows. *Journal of Cleaner Production 250*, 11pp.

15

LIVING IN CITIES

15.1 Cities and growth

The point of writing this book was to investigate whether the move to increased urbanisation of the human population is a good thing. What emerges is it is not urbanisation itself that is the problem but the reason behind the wish to live in cities, which is bound up in economic growth. This raises the issue of what economic growth is for, since currently it seems to be leading to failure to deal with issues such as social inequality and climate change. In 1964 the economist James Tobin noted that economic growth had become important in governments and government policies "both in the USA and abroad". More intriguingly he also wrote the following (Tobin, 1964):

> In essence the question of growth is nothing new, but a new disguise for an age-old issue, one which has always intrigued and preoccupied economists: the present versus the future. How should society divide its resources between current needs and pleasures and those of next year, next decade, next generation?

The idea of investing in the present for the future is as familiar as spending time cooking the evening meal or building a house. For governments this means the same thing; governments build infrastructure, such as roads, that causes inconvenience now but that they think will be time saving and convenient in the future. The current problem is that Tobin's words can be read in a different way. As the Brundtland Report (World Commission on Environment and Development, 1987:43) suggested, current generations should be saving resources for future generations and taking steps to ensure that there is no climate change catastrophe. Given that the human population is already in earth overshoot and that living in

DOI: 10.4324/9781003045847-17

cities tends to increase rather than decrease the use of resources, it seems that a new city model is required, rather than more of the same, if this is where people are to live.

15.2 Voting for a new city model

As discussed in Chapter 8, good intentions are not difficult to put into place. Achieving actions for the common good, such as dramatic reductions in CO_2 emissions, which in turn require lifestyle changes, is much harder. The democratic system was set up so that decisions which affect all can have input from all concerned. In a small group such a system can work well, but in large societies, a vote every few years to elect a government that might or might not fulfil election promises seems a long way from the original intention of a democracy. It also seems that dramatic change occurs outside of the democratic process, as demonstrated by the government-initiated lockdowns in the face of the threat to life from COVID-19. The great majority accepted this curtailment of individual freedom in favour of the common good, as also shown with the acceptance of mass vaccination programmes. Big changes are possible when the threat is real. Unfortunately, asking people to give up their cars to prevent significant global warming is not yet part of any elected government's strategy because the threat is not perceived as imminent, although it will affect our children and their children.

It also seems from the research that went into this book that no regime is performing dramatically differently from another when it comes to dealing with environmental issues. This links back to the fact that all governments pursue economic growth. Back in 1995, Grossman and Kreuger claimed there was "no evidence that environmental quality deteriorates steadily with economic growth. Rather, for most indicators, economic growth brings an initial phase of deterioration followed by a subsequent phase of improvement." If this is the case why should a government give up the idea of economic growth, which brings with it more employment and more taxes, so more money for supporting government policies? The reality is that climate change goes well beyond the purely local indicators of environmental quality used to justify economic growth by Grossman and Kreuger (1995), which included local pollutants but not global CO_2. Economic growth links directly to climate change due to the strong connection between economic growth and energy consumption, and hence climate change (Akram, 2013).

15.3 What a new city model might be

The city of the future will not be based on people coming together to make money. It could be based on people coming together for innovation in the arts and for continued research into a way of life that can be sustained over the generations. Living at high densities will still need hinterlands to supply citizens with food and other essential resources such as clothes and energy. To keep this area as small as possible, the city will not be a place where the aim is to have more, but only to have sufficient. Coming together in the city will be for the purpose of saving land to be put to better uses, such as agriculture and areas set aside for species other than human beings, rather than for degradation by buildings and infrastructure.

Physically the city will need to save as much energy as possible in its built environment, which must be designed to last as long as possible with simple maintenance. There will be no cars, the age of the car is over, the main modes of transport will be walking and cycling with electric mass transit systems, perhaps underground, for longer travel within the city. There will not be airports, with international trade confined to rail and shipping. Buildings will be walk-up apartments rather than expensive high-rise structures. As demonstrated by earlier discussions in this book, these are the areas where big savings in greenhouse gas emissions could be made.

15.4 How to get there from here—A to B or A to C?

The original intention of this book was to compare cities from around the world with different governance structures and forms to see if something could be learned about what was already working to reduce the environmental impact of living in cities. This intention soon became unrealisable because of the lack of comparable data, something also noted by Grossman and Krueger in their 1995 study. Perhaps the first step in knowing how to develop a new city model for a burgeoning human population in a time of climate change is to know where we currently stand in terms of energy use, whether this is energy used for transport or home space conditioning or energy embodied in the built environment and in goods and services. With this information, setting policies to reduce environmental impact would have measurable outcomes, with the aim of going from A—where we Are, to B—where we want to Be, in terms of environmental impact reduction.

All the data we have found in the research for this book points to the fact that none of the current political and economic systems are taking us in the right direction. Unfortunately, the two most highly favoured systems for organising the world, democracy and capitalism, clearly seem to be taking us, not from A to B, but to places beginning with C—climate change, collapse and catastrophe.

References

Akram, N. (2013). Is Climate Change Hindering Economic Growth of Asian Economies? *Asia-Pacific Development Journal 19*(2), 1–18.

Grossman, GM. and Krueger, A.B. (1995). Economic Growth and the Environment. *Quarterly Journal of Economics 110*(2), 353–377.

Tobin, J. (1964). Economic Growth as an Objective of Government Policy. *The American Economic Review 54*(3), 1–20.

World Commission on Environment and Development (1987). *Our Common Future.* Oxford: Oxford University Press.

INDEX

For Product Safety Concerns and Information please contact our EU representative GPSR@taylorandfrancis.com
Taylor & Francis Verlag GmbH, Kaufingerstraße 24, 80331 München, Germany

www.ingramcontent.com/pod-product-compliance
Ingram Content Group UK Ltd.
Pitfield, Milton Keynes, MK11 3LW, UK
UKHW020150200926
475702UK00016B/2637
9780367493424